SECURING UTILITY AND ENERGY INFRASTRUCTURES

SECURING UTILITY AND ENERGY INFRASTRUCTURES

DR. LARRY NESS

**WILEY-
INTERSCIENCE**

A JOHN WILEY & SONS, INC., PUBLICATION

Published by John Wiley & Sons, Inc., Hoboken, New Jersey.
Published simultaneously in Canada.

For general information on our other products and services or for technical support, please contact our Customer Care Department within the United States at (800) 762-2974, outside the United States at (317) 572-3993 or fax (317) 572-4002.

Wiley also publishes its books in a variety of electronic formats. Some content that appears in print may not be available in electronic format. For information about Wiley products, visit our web site at www.wiley.com.

Library of Congress Cataloging-in-Publication Data:

Ness, Larry.
   Securing utility and energy infrastructures / Larry Ness.
      p. cm.
   Includes index.
   ISBN-13 978-0-471-70525-3
   ISBN-10 0-471-70525-X
   1. Public utilities—Security measures—United States.   2. Energy industries—Security measures—United States.   I. Title.
HD2766.N455 2006
333.79068'4—dc22   2006007170

Printed in the United States of America.

10 9 8 7 6 5 4 3 2 1

Deepest appreciation is expressed to my daughter, Ashley Ness,
who made, by far, the greatest sacrifices
in assisting me during this entire process.

ACKNOWLEDGMENTS

The author wishes to express appreciation to a number of professionals who greatly aided in this endeavor.

Special appreciation is extended to Mr. Robert Wood for his continuous encouragement and words of support. Also my appreciation goes to Dr. Alex Hapka for his guidance and encouragement during this project.

The following are national professionals who generously supplied articles, interviews, and many hours of coaching and advice. This project would not have been possible without their unselfish cooperation and assistance.

Mr. Pete Arnold
Mr. Richard Guilmette
Mr. Joseph Weiss
Mr. Robert Huber
Mr. Robert Schainker
Mr. Brian Wolf

Mr. Dave Juring
Mr. Raymond Humphrey
Mr. Doug McQueen
Mr. Mike Ness
Mrs. Ann Diamond
Mr. Brian Walshe

Mr. Frank Earl
Mr. Bob Temple
Dr. Richard Harper
Mr. Michael Steinle
Mr. Joe Gulienello
Mr. Massout Amin

ABOUT THE AUTHOR

Larry Ness, Ph.D., is a senior security executive and president/owner of Ness Group International. Ness Group International is a Dallas, Texas-based "licensed & insured" global security consulting and investigative group that specializes in the utility/energy industry.

Mr. Ness is a former member of the U.S. Secret Service, where he was also a faculty member with the Federal Law Enforcement Training Center. He served as a corporate security director for a major utility, and was Executive Director for the Electric Power Research Institute, Dallas, Texas.

Mr. Ness served as Chairman of the North Dakota Parole Board, was past member of the North Dakota Pardon Board, and was President of the Board, Heartview Foundation, a drug and alcohol treatment facility.

He is a seasoned speaker and has given numerous presentations on utility/energy security. He has also published articles on security in national energy publications'

Mr. Ness holds a Ph.D. in Criminal Justice Management, MS in Criminal Justice Administration and BS in Law Enforcement. He has also been active over the years as a member of the American Society for Industrial Security, International Association of Chiefs of Police and part-time Adjunct Professor in Criminal Justice for two colleges.

CONTENTS

Chapter 1

INADEQUATE SECURITY TODAY

The U.S. National Academy of Engineering has evaluated all the engineering achievements in the United States for the past 100 years and has determined that "the vast networks of electrification are the greatest achievement of the 20th century." Building and maintaining our nation's electric grid are the result of a public and private partnership that has worked for almost a century.

Think about how your life is dependent on the electric grid. Almost every aspect of material existence relies directly upon electricity for light, heat, refrigeration, air conditioning, communication, financial transactions, and so much more. Transportation, food distribution, schools, hospitals, and government services all rely on electricity.

Now think what life would be like without the electric grid. With no electricity. We would be thrust back 150 to 200 years. There would be no computers, no television, no landline or wireless phones, no cars or trucks, no planes, no banks, and no stock markets. We would revert to an agrarian economy, but we wouldn't have nearly enough water to drink, food to eat, or wood or coal to burn in stoves or fireplaces.

The electric grid is just one of the utility/energy infrastructures that are fundamental to our way of life. Others include natural gas and gasoline; the chemical, nuclear, and telecommunications industries; trucking and pipeline transportation systems; and water utilities.

Our utility/energy infrastructures—the delivery systems—are inextricably intertwined with our lives. We are so dependent upon them that, for most of us, they are invisible. But they are extremely vulnerable to attack by terrorists and others who would harm us. In *Securing Utility/Energy Infrastructures,* I will detail what has been done to improve our security, what needs to be done, why it must be done, and innovative ways to pay for it.

My purpose in writing this book is to help create a more secure environment in the United States. The good news is that we can do this if we make security a higher priority. We have excellent people throughout the utility/energy industry, as well as most of the necessary knowledge and much of the technology and equipment. The money to pay for improved security, often cited as the excuse not to do much or anything, will not be a problem. I promise you that.

The big questions are: Do we have the will? Will we be proactive in improving our security? Or will we wait until we sustain another terrorist event and then react with tough words and recrimination?

HISTORY OF THE UTILITY/ENERGY INDUSTRY

The infrastructures of the utility/energy industry are 40 to 50 years old. Many of them need attention because of their aged and in some cases deteriorating condition. But even if our infrastructures were continuously refurbished, they were not built with security as a top priority. The traditional security measures are "guns, gates, and guards." At many remote electric sub-stations, even these standards are minimally maintained. I have seen rudimentary perimeter fences that are not carefully monitored and too few poorly trained guards who do not adequately cover all entrances and exits.

Cyber security at too many of our utility/energy installations is elementary at best. One member of senior management at a large electricity-generating plant recently told me, "Of course we have cyber security. We have all kinds of email protection."

New employees and vendors are not thoroughly screened, and internal structures have not been built to protect against an attack by even a two-passenger plane. This lack of security is not news to most people within the utility/energy industry. They are aware that, when it comes to securing our nation's utility/energy infrastructures, too little has been or is being done in Washington. And too little has been or is being done in the great majority of state and local governments.

FEDERAL LEGISLATIVE ISSUES

One of the obvious places to address improved security for our nation's utility/energy infrastructures is in Congress. The much debated and long-delayed 816-page Energy Bill has languished in the House and Senate for several years. Legislators can't separate the reliability issues from the highly politicized battles over drilling for oil in Alaskan wildlife refuges, the regulation of carbon-dioxide emissions, a giant mall that wants tax-exempt green bonds, the attempt to eliminate tariffs on Chinese ceiling fans, and so much more.

Yet there are a great variety of tax incentives for utility/energy companies to put security in place, especially cyber-security technologies that can help to prevent another blackout such as the one that occurred in the Northeast and parts of Canada on August 14-15, 2003. In Congress and in "The 9/11 Commission Report," there is little focus on improving the security of America's utility/energy industry.

DEREGULATION OF THE UTILITY INDUSTRY

The age of our utility/energy infrastructures, their lack of adequate security, and the insufficient focus of Congress and other government agencies on these vital infrastructures are compounded by the deregulation of the utility/energy industry. According to "Electricity Sector Framework for the Future," published by the Electric Power Research Institute (EPRI) in 2004, "Electricity service is, by its nature, an extraordinarily capital- and technology-intensive, politically constrained enterprise, and coherent leadership is needed to break today's conflict logjam. Since the open access order in 1992, the institutional structure for the electricity sector has been dismantled, but it has not been replaced by an alternative structure with coherent institutions and rules. As a result, elements of the diverse electricity sector are already in crisis, and the impacts have been spreading. The investor-owned utilities, in particular, are laboring under an inconsistent and conflicting set of regulations."

According to this EPRI report, "The public power sector, while relatively stable, is now also being impacted by the ripple effects of restructuring. Market reforms—worthy, limited experiments that have shown mixed success—have resulted in rules that differ from state to state, and in many cases, from utility to utility within a state. At the federal level, open access has been mandated but without clear direction on how it is to be implemented. Fluid environmental policies and proposals have simply added to the uncertainty in the electricity sector, even though many of the proposals are intended to reduce uncertainty."

Before deregulation, electricity was most often generated close to where it was used. After deregulation, electricity is typically transmitted from power plants in one area of the nation on high-voltage lines to commercial users and consumers who live hundreds of miles away. That exposes us to interruptions over greater distances.

The EPRI report continues, "Any of these problems alone might have been manageable. But the simultaneous convergence of several independent issues has caused serious turmoil in the business aspects of the electricity sector. Wholesale markets are increasingly thwarted by the inability of an aging U.S. power delivery system to support transactions. Further expansion of retail deregulation has essentially come to a full stop. Credit markets have shut out nearly all of the high-risk 'merchant energy' companies, whose business in recent years has turned from boom to bust. Other industry members have seen their credit ratings drop, and financing costs for the industry have risen dramatically. The impact of these difficulties is an inability to plan, an unwillingness to invest, and a stalemate in strategy for achieving a way out of the current dilemma.

"As a consequence of these issues, our nation's utility and energy companies have inadequate physical and cyber security, and most of their security personnel are not screened or trained well enough. The utility executives who want to make improvements in these areas most often believe the changes would drive up their cost of power, making their service noncompetitive."

The majority of utility executives I have talked with don't envision that their facilities and infrastructures face a reliability issue without increased security. Instead, day in and day out, they say that their service is performing well without the need for additional safeguards.

Nor do these executives think strategically about expanding their vital services to new profit centers. A few pages above, I promised to tell you how we as a nation can afford to secure our utility/energy infrastructures. Here's an introduction to this vital subject: I envision our nation's utility/energy companies providing their traditional services while also expanding to offer a variety of security services for their commercial and consumer customers. I will discuss this subject in greater detail in Chapter

Figure 1.1 Substation control room

12. Our nation's utility/energy infrastructures can be made far more secure while they simultaneously provide increased services that will help to pay for this security.

THE ROLE OF FEDERAL AND STATE UTILITY REGULATORS

"Trees or terrorists, the power grid will go down again," states Robert Schainker of EPRI, the research arm of the nation's power utilities. This generally accepted view is one of the reasons that the Department of Energy (DOE) and the Federal Energy Regulatory Commission (FERC) have begun to develop procedures that require the utility industry to help secure the nation's electric grid. Following the 2003 Northeast blackout, for example, Congress appropriated an additional $5 million for fiscal 2004 alone so that FERC could better address grid reliability issues.

Some key accomplishments of FERC's reliability effort during 2004 include:

- Urging the industry-led North American Electric Reliability Council (NERC) to adopt tougher, clearer, and more enforceable reliability standards to replace the current voluntary reliability guidelines.
- Working with NERC to perform reliability audits.
- Undertaking research studies on reliability-critical issues, including replacement transformers, grid operator training, System Control and Data Acquisition (SCADA) vulnerability to cyber attack, as well as the potential impact of natural-gas-pipeline disruption of gas supplies and electricity production.
- Working with partners including the Canadian government, state electricity regulators, and the U.S. Nuclear Regulatory Commission to develop long-term institutional solutions for grid-reliability challenges.

Officials at FERC fully understand the challenges facing our nation's electric grid. Here's a summary statement of a report that FERC has sent to Congress:

"The North American electricity grid faces a challenge–it uses some technologies invented in the early 1900s and facilities built primarily in the mid-20th century to serve an economy with electricity demands and needs of the 21st century. Today's needs reflect unprecedented levels of demand. In the United States, electricity demand has grown by 2.1 percent annually over the past ten years. Much of that electricity is used to power appliances such as computers, medical equipment, and industrial machinery that can be harmed by power quality fluctuations or even momentary power interruptions.

"But while there has been extensive investment in generation plants to meet the growing thirst for electricity, there has been little corresponding investment in transmission lines to connect those loads to the plants—between 1986 and 2002, peak demand across the United States grew by 26 percent and U.S. electric generating capacity grew by 22 percent, but U.S. transmission capacity grew little beyond the interconnection of new power plants.

"As load grows, the increased power flows across the lines from producers to users are causing increased congestion on the lines and increasing the difficulty of operating the system safely and reliably. One analysis found that between 1982 and 2002, normalized transmission capacity declined at a rate of 1.5 percent per year, although electricity sales nearly doubled, and concluded: Most of the recent and planned investment in transmission facilities is intended to solve local reliability problems and serve growing loads in large population centers. Few projects cross utility or regional boundaries and are planned to move large blocks of low-cost power long distances to support large regional wholesale electricity markets. Thus, many opportunities to lower consumer power costs will be forgone because of insufficient transmission capacity."

To deal with part of the reliability issue, FERC is developing orders that will require transmission-system operators to report violations of the industry's power-grid reliability standards. Keep in mind, however, that some 40 percent of the electricity generated nationally is provided by companies covered only by voluntary guidelines.

A number of states are not waiting for agencies of the federal government to act by dictating policy and providing dollars to secure the industry. Instead, California, Iowa, New Jersey, and New York are among the states that are stepping up to the plate to develop security standards and allow rate increases to pay for them.

POST 9/11 INFORMATION SOURCES

According to the 9/11 Commission Report, "The 9/11 experience shows that terrorists study and exploit America's vulnerabilities." Here and elsewhere in this book, I will only disclose problems that are already public knowledge. I need to call attention to some ways that we are most vulnerable, yet of course I must avoid disclosing confidential information. A team of national experts, therefore, has reviewed this manuscript before publication to help ensure that no information provided will give terrorists any advantage.

I am profoundly proud of the history of our nation's utility/energy industry. Our electric, natural-gas, oil-refining, chemical and biochemical, water, nuclear, and telecommunications industries have served America's needs during peace and war, during boom times and depressions. Decade after decade, when earthquakes, hurricanes, major fires, and other disasters have occurred, we have seen consistent cooperation from unaffected utilities that have rushed to help meet the energy needs of people, businesses, and governments in disaster areas.

When the infrastructures of these industries have been impacted, industry representatives from near and far have always come to help. The 2003 blackout is a particularly relevant example. It affected eight northeastern states and parts of Canada for two or more days during mid-August. In fact, the worst blackout in North American history struck without warning shortly after 4 p.m. on August 14, when 61,800 megawatts of electricity went offline, cutting power to Michigan, Ohio, Pennsylvania, New York, Massachusetts, Vermont, Connecticut, New Jersey, and Ontario, Canada. New York, Detroit, and other major cities broiled in 90-degree heat.

People realized that, without electricity, they couldn't get gasoline, use ATMs, or drink non-boiled water. The estimated cost of the blackout in the United States was at least $6 billion, with some estimates ranging between $7 and $10 billion.

I've talked with several people in the Northeast who lived through this blackout. Some had to climb 20 or more stories to get to their hot apartments because neither the elevators nor their air conditioners worked. Perishable foods were eaten fast or thrown away. Toilets were flushed only once when absolutely necessary, then not flushed again. Only bottled water was drinkable.

Ryan Goodfellow is a college student who was home during those summer days with his mother, father, and two younger sisters. They live 20 minutes from Syracuse, New York. "It was a life-changing experience," he says about the blackout. "We had no power and no water, so we didn't open the refrigerator at all. For the first day, I ate corn chips. That got a little tiring, so on day two, we got into the car, picked up grandma, and headed into Syracuse, where we stayed at a hotel that had a generator."

In Michigan as well as other regions, traffic lights didn't function. Systems at the Michigan Intelligent Transportation System Center, including video cameras and the Center's website, went down, and the Detroit(Windsor Tunnel was shut because the ventilation systems didn't operate. For most people, the blackout was a major inconvenience, with tens of thousands of people unable to get home for the night or delayed hours in that process. In New York City, there was an eerie feeling about the possibility that it was a terrorist incident or maybe a trial run.

Several months afterward, the Final Report of the U.S./Canada Power System Outage Task Force was issued. It listed four causes of the blackout:

- Inadequate system understanding
- Inadequate situational awareness
- Inadequate tree trimming
- Inadequate reliability-coordinator diagnostic support

The Report also identified seven violations of the voluntary reliability standards administered by the North American Electric Reliability Council (NERC).

To the best of my knowledge, there are no valid studies that prove that it was a terrorist event. Almost surely, it wasn't. Many commentators have asked, however, whether it was an invitation to terrorists. Is the havoc caused by this blackout too tempting for terrorists with advanced computer skills to resist?

Whatever the answers, it is not in our nation's interest to sit back inactively. The 2003 blackout clearly shows that some of the vulnerabilities of our nation's utility/energy infrastructure must be addressed. There were also massive power outages in the United Kingdom and Italy in 2003 and in Greece in 2004.

In all cases, some remedial measures have been applied to correct problems. But in the great majority, the basic vulnerabilities of our nation's and the world's utility and energy infrastructures are not being addressed.

A study produced in 2004 by the National Academy of Sciences identifies common weaknesses in our electric grid that were evident in the 2003 blackout and in previous blackouts. According to this study, there is much room for improvement:

- Monitoring of the power grid is sparse, and even limited data is not shared among power companies.
- Monitoring of the power system is inadequate everywhere, both within regions and between them.

- Industry standards are lax; for example, vegetation under transmission lines is trimmed only every five years.
- Operators are not trained routinely with realistic simulations that would enable them to practice dealing with the precursors to cascading failures and the management of large-scale emergencies.
- Power companies have widely varying levels of equipment, data, and training. Some companies can interrupt power to customers quickly during an emergency, whereas others are nearly helpless.
- Decades-old recommendations to display data in a form that makes it easy to see the extent of a problem have been ignored.

A little more than one year after the blackout, four hurricanes that struck Florida wreaked hardship on millions of people. Of course there is no way to prevent these natural disasters, and yet the deprivations they caused have some similarities with the consequences of attacks on our utility/energy infrastructures. Ted Wilhite has lived in Florida since 1944. Never before had he been subjected to four hurricanes and one tropical storm within a matter of weeks.

His two-story house on an island in Tampa Bay was built on piers 1.5 feet above the ground. "We evacuated for Hurricane Charley, which gave us wind and lots of rain plus limbs down and flooding. We returned home and along came Francis from the other side of the state. We didn't need to evacuate for Francis, but we were without electricity for five days. Lots of trees and limbs were down. Jeanne was the last to affect us. We had 90-mile-per-hour winds and no electricity for another two days.

"We lost all the food we couldn't consume fast. We had friends staying with us who had evacuated from Melbourne, and we watched a TV that ran on batteries. Soon we needed batteries, ice, newspapers, and a liquor store. I found everything but the D batteries. We put the ice in coolers.

"We barbecued as long as we could, then we ate a lot of canned goods. When the electricity came back the first time, my wife said I shouldn't buy much food at the store. I didn't, and she was right. Other storms that knocked out the electricity again would have made us throw that food out, too. And then when the electricity was finally restored, three houses in our area caught fire. Undetected internal damage was ignited by frayed power lines. "Keep in mind that there were people in the state who were without electricity for weeks. When it's hot and humid with lots of mosquitoes, it can be tough. The heat can be a killer," Ted Wilhite concludes.

"Where do you begin?" asks Kevin Brown of Clearwater, Florida, after sustaining the fourth hurricane, Ivan. "You want to get back everything, make it's like it was, and you can't. Ever again." His wife wants to move, but he's not sure. "Sometimes I think I want to build back up again, and then I think, what's going to happen next time?"

Vicki Woolford and her husband, Alan, were unable to return to work for weeks. For income she did housekeeping at a local motel. They spent part of every day cleaning up what's left of their house near Perdido Bay. They plan to rent a doublewide mobile home and live nearby while their house is rebuilt. The new version will sit on stilts.

The absence of electric, gasoline, and telephone utilities was among the many hardships causes by the hurricanes of 2004 in Florida. The high winds and water damage caused the majority of the damage. My point in citing these events in the Northeast during August 2003 and Florida during late summer 2004 is that they provide some precedent for Americans doing without what we have come to believe are basic necessities. The lack of adequate security at the nation's utility/energy infrastructures could result in these same problems and much worse.

Please understand that the very competent people who manage our nation's utility/energy infrastructures often act upon important priorities that have no relationship to security issues. They have complex organizations to run. Some of these entities are for-profit, some are nonprofit,

Figure 1.2 Substation

most are small and rural, and a few are huge and urban. Some have the financial resources to make necessary improvements; others have funding to do what's necessary and little more.

But whatever the status or location of these facilities, I have one question: What would happen if they and others were knocked out of commission for six months? It's one thing to have a blackout for about 48 or 72 hours—or hurricanes that come time and again over a period of weeks. But what would it be like if terrorists attacked the nation's electric grid and crippled it for half a year? And what if that long blackout occurred in the northern half of America during the winter?

Here's what people who have been without electricity for just two, four, or more days said. "People would die," said Judy Wynbrandt, who has lived in Manhattan for 25 years. "They wouldn't have food, water, or heat. They'd die." Other people I've talked with have said simply, "We'd make do. Trucks would bring in food, natural gas, gasoline, and home heating oil."

But how would we do that? What would we use to pump fuel into the trucks? There would be no traffic lights, no streetlights, and tens of thousands, even millions, of people would become desperate for heat, food, and water. Crime would run rampant. People would fight for survival, starve, and freeze to death.

RISK ASSESSMENT VERSUS NEEDS ASSESSMENT

There's a good chance that you're asking, "What's the probability that any of our utility/energy infrastructures could be out of commission for up to six months?"

Nobody knows. Some think tanks study these issues and present their findings, but the end result is that nobody knows. We do not have the intelligence factors to quantify these metrics, so we can't talk realistically about probability.

There are, however, other prudent responses. According to Massachusetts-based Raymond Humphrey, President of Humphrey & Company and board-certified in security management, there's much we can do. "To minimize our risks, we can talk about risk assessment and compare it with needs assessment," he says. "We can examine what's the most critical aspect that we can't afford to lose and then take actions to protect what is most vulnerable."

We all do this as we protect our personal assets against disaster via homeowner's insurance, appliance warranty programs, bumper-to-bumper coverage on our cars and trucks, and service contracts on our computers.

As utility customers, we need to think of ourselves as being "agents of protection." We can be a driving force, from both a customer and shareholder point of view, to motivate and help utility companies implement appropriate security programs to protect our personal assets—so many of which are intertwined with electricity and natural gas.

A similar response is offered by The Guilmette Group, an international security consulting firm based in Franklin, Massachusetts. Its principal, Richard Guilmette, states, "For our clients with questions about probability, we use the Risk of Loss Quadrant (Figure 1.3) to conceptualize the elements of threat and vulnerability. You can mitigate vulnerability when you understand how you are vulnerable. Know your weaknesses and learn what needs to be done to reduce the exposure."

"Threat, on the other hand, is more difficult to pin down," Dick continues. "We can't necessarily know the intentions, capabilities, or capacities of those who would harm us by exploiting our weaknesses. Good intelligence is a step in the right direction. Knowledge of and vigilance about the threats help to identify their characteristics. But in the final analysis, it's best to take proactive measures that direct terrorists to softer targets in other countries far from America. We want them to consider the risk too great at our hard targets. That is likely to be the most expedient option." we can use the Guilmette Risk of Loss Quadrant to better understand any important asset. Let's take our nation's electric grid and view the risk of loss to that asset.

We've already focused on the vital use of electricity we Americans make every day. There is probably no material asset of greater value.

A cascading failure could occur from one region of the country to another, from power junctions and power stations with crossover capabilities through critical junctions, which are also called critical nodes. The 2003 blackout cascaded from a local to a regional meltdown when all the backup systems failed. It was a wakeup call to look at systems that have been in place for decades. So the risk of loss from electric power grids can vary from significant to great to catastrophic.

What is the vulnerability, which is expressed on the V line, of this asset? Look at the vulnerabilities: What is susceptible? Where can we get hit? There are approximately 5,800 power-generating plants and 10,200 substations around the country, and there is no such thing as 100 percent protection.

"Criticality is the main issue," says Ray Humphrey. "Where would the loss be greatest? That's what must be the best protected, because that's where we're most vulnerable." Those who would harm us would get the maximum impact by causing damage to our electric grid that would affect our financial communities, communications and transportation systems, and alarm systems in large institutions and also disrupt the ability of government and public safety to respond. Says Dick Guilmette, "A coordinated attack at multiple points of vulnerability would lead to cascading problems. An attack on two fronts would include a cyber attack to get into the SCADA network itself, which damages the SCADA structure, while co-conspirators would conduct physical attacks in different parts of the country where there are gateway nodes."

We are most vulnerable at those gateway nodes, where damage could cause cascading blackouts across the nation. The threat level, however, is hard to gauge until we get much greater cooperation between the private and public sectors. We need better intelligence from corporate security forces working in conjunction with local and state law enforcement, government intelligence, and law-enforcement associations, such as the American Society for Industrial Security (ASIS).

Only through greater coordination of our various law enforcement capabilities can we gather the intelligence that will accurately determine the changing threat levels.

Risk of Loss Defined
$$R_L = f(T, V)$$

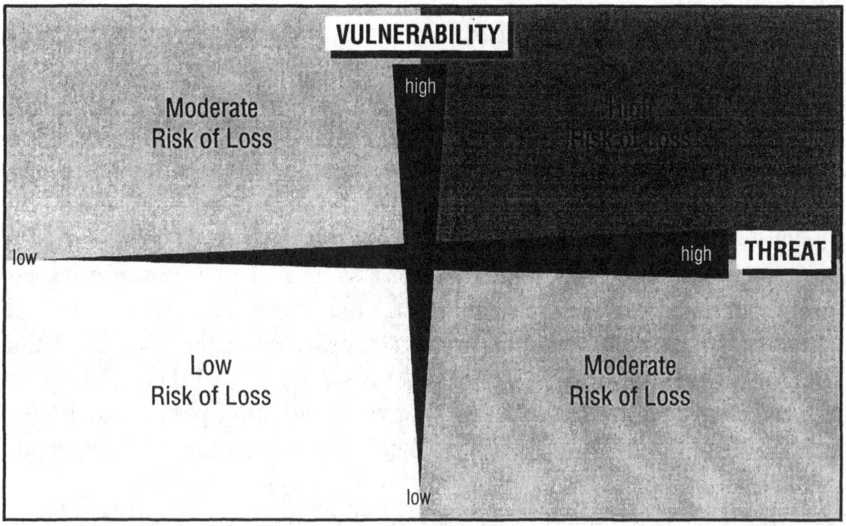

Figure 1.3 The Guilmette Risk of Loss Quadrant helps determine where the risk of loss is greatest.

Note that some vulnerabilities and their consequences aren't completely obvious. For example, the electrical system within a nuclear power plant would immediately cause problems for the generators.

The threat line is for people who want to do harm to us. They are different from hurricanes, earthquakes, fires, and other such disasters. The latter are regionally based and are in sharp contrast to attacks from people who could bring down the SCADA system.

SCADA stands for Supervisory Control and Data Acquisition Systems. It supervises and controls our power grids, gas pipelines, and oil refineries. Although we have significant safeguards, SCADA systems can be accessed and manipulated remotely. They are a potential prime target for terrorist groups, which could get inside them and inflict physical damage on the nation's utility/energy infrastructures. "Because this is a human-generated event," Dick Guilmette states, "the results are not as predictable as a natural disaster. We need to understand the nature of the threat, if it's from al Qaeda, a super-nationalist organization, or a disgruntled employee."

The threat line in his quadrant goes from an unsophisticated instigator to one who is sophisticated, from 1 to 5 and then from 6 to 10. The unsophisticated perpetrator is on the outside looking in, while the sophisticated cyber attacker understands the SCADA system from the inside. He or she most likely was trusted by officials at one point, then turned against them.

According to Ray Humphrey, who uses the Guilmette Risk of Loss Quadrant, "Our highly mobile workforce results in transparency of loyalties. Information technology (IT) employees move from company to company, even from one country to another. In addition, the threat line is blurred between unsophisticated and sophisticated individuals as IT information is passed among people and through the Internet."

Personnel at utility/energy infrastructure locations teach a seemingly loyal workforce proprietary information, and then employees move on. We must make sure that knowledge is compartmentalized so people at lower and middle levels only know portions of the confidential information. We must continuously do background checks.

"We know that we absolutely need to protect our nation in terms of energy transmission," Dick Guilmette states. "Given that our resources are finite, we must protect against cascading effects that could take out the national electric grid." The security function in a profitmaking or publicly-held utility/energy infrastructure is competing against many other budgetary demands. Management of the facility needs to weigh the threats against the vulnerability in the context of the financial resources. In the end, many managers whose facilities are not part of the critical nodes look to expert advice beyond risk-assessment models and methodologies to conduct needs assessment specifically related to individual infrastructures in particular locations. That is often the most reliable way to determine what is a reasonable threat.

There are other important responses to the issue of risk assessment versus needs assessment that relate to a new way of thinking. I'll describe them and much more in Chapter 2.

Chapter 2

THE ECONOMIC IMPACT OF A TERRORIST ATTACK

We can be a safer, more secure country if we expand our thinking in ways that some might never have considered. In this chapter, I will show you the economic costs of 9/11, and I'll compare that tragic day with other national and international disasters. I'll also ask you to help prevent a future we hope never occurs by contemplating what could go very wrong. I believe that's the best way for us to begin thinking and acting now to prevent major catastrophes.

UTILITY INDUSTRY–PRIME TARGETS FOR TERRORISTS

There are a variety of significant threats to our nation's utility/energy infrastructures. These threats come from several sources, although the most damaging would be from terrorists:

- New hires: It is extremely important to prescreen new employees by running background criminal checks to ensure that they are who they say they are.

- Vendors/contractors: Because of deregulation in the utility/energy industry, companies are downsizing and contracting out services. Due-diligence checks that include terrorism affiliations need to be done on contractors and security checks need to be conducted on their employees.

- Disgruntled employees: They could be knowledgeable about particular energy facilities, and they could get past gates and guards to damage infrastructures. If they also are familiar with cyberspace, disgruntled employees can hack into a utility's computers to cause damage. An employee who was fired from his job at a hotel sought revenge that would offend lots of people. He invaded a nearby utility's computer and caused tons of raw sewage to flow on the hotel's grounds.

- Computer hackers: Some seek to infiltrate large systems just to prove that they are more skilled than the system's designers. Others could be funded by angry stockholders who face major losses if corporate fraud were to occur at any utility/energy companies.

- Game players: Seemingly innocent players of computer games have broken through firewalls to access the more powerful computers in energy facilities. In some cases, once inside the power plant's computers, these hackers have found a new game: shutting down utility boilers. Why do they do it? Just to show they can.

- Islamic terrorists: They pose by far the greatest threats. Just as the perpetrators of the 9/11 hijackings went to flight school to learn how to navigate commercial airliners, today thousands of computer-science students from countries that harbor terrorists are enrolled in American universities. Of course this doesn't necessarily mean that any or many of these students will become terrorists. But we should definitely be alert and far better prepared for actions that terrorists might take.

Except when there's damage from a hurricane or tornado or during disruptions such as blackouts, we take our nation's utility/energy infrastructures for granted. Yet these most fundamental infrastructures in our nation carry electricity throughout the country from our electric-power systems. After 9/11 we significantly increased airport and seaport security. That was and continues to be necessary, but it is not sufficient.

Security at our air and seaports is dependent upon electricity. Of course, these security operations have backup energy generators, but they only operate for a specified amount of time before they need to be replenished. I also wonder how many of these backup systems are checked regularly to ensure they that will start when needed.

Beyond the security systems that help screen airplane passengers, we have approximately 6,000 commercial flights per day all over the United States. The air-traffic-control systems that manage all these flights depend upon electricity. We and every other nation in the world are vulnerable to attacks on our power-generating infrastructures.

But if you think that's troubling, let me be straightforward: The situation is far worse than simply being vulnerable. Our nation's utility/energy infrastructure can be used against us. The threat is similar to the commercial aircraft that the 9/11 terrorists transformed into bombs that crashed into our buildings.

MAJOR TYPES OF THREATS

There are five major types of threat to our nation's utility/energy infrastructures. The first threat is a direct attack on the infrastructures. These assaults could be on the ground, using suicide bombers and trucks laden with explosives, or through the air, even using single-engine planes that fly from small airports with far less security than the major hubs. A direct attack could take down a local area, a vast region, or a major portion of the country, even the entire electric grid.

Use the Infrastructure as a Weapon Against Us

The second threat is to transform the infrastructures into weapons used against us. For example, terrorists could dump chemical or biological materials into a power plant's operating system, which could result in contaminating a large area via the cooling towers.

Create Power Surges

The third is to create power surges through the grid. These surges can ruin computers and cause major damage to telecommunications systems and related technologies. If these systems are interrupted, communications will be lost and financial transactions of all sorts will stop.

Cyber Attack

The fourth is a cyber attack. Hackers located anywhere in the world could break into the computer systems that control the utility/energy

Ness Group International
Professional Loss/Risk Consulting

Larry Ness
President/Owner
Ness Group International – Dallas, TX
WWW.NESSGROUP.COM
LNESS@NESSGROUP.COM
214.415.9687 Office

EMPLOYEE BACKGROUND SCREENING POLICY

**Effective August 1, 2002, _____will
introduce new policy guidelines that will identify the process for
verifying information given to _____ by
prospective employees.**

The purpose of the policy is to provide management with guidelines to
insure consistency in the verification process and to insure compliance with
the provisions of the Federal Fair Credit Reporting Act and other applicable
State and Federal Guidelines.

The following is an overview of the policy and process, which will be
administered and managed by the Corporate Security Director.

All applicants who are interviewed, either in person or via telephone, must
be informed at the time of the interview that _____ will
conduct a background check **after an offer has been tendered and
accepted.**

The background check will consist of the following:

- Verification of the information given to _____ on the
  applicant's resume and/or employment application, e.g., work history and
  education,
- Social Security Number verification,
- Criminal history check and,
- Driving record verification (Applicable to those who are responsible for
  driving company owned or leased vehicles. This does not apply to vehicle
  rental while on business travel).

Applicants must also be informed that, if selected, they will be required at
the time of hire to sign a **Notice and Consent form,** which authorizes
_____to conduct the background check, and

page 1 of 3

Figure 2.1 Employee background screening policy *(continued on next page)*

Release of Liability form.

The Notice and Consent and release of liability forms need only be given to applicants who **are offered and accept** a position with _____. To clarify, all applicants who are interviewed must be notified that _____will conduct background checks if and when they receive and accept an offer of employment, but only those for whom an offer is tendered and accepted will Notice and Consent and Release of Liability forms be presented for signature.

After accepting an offer for employment, the New Hire will receive the following documents:

- Notice and Consent (Must be signed)
- Release From Liability (Must be signed)
- Prescribed Summary of Consumer Rights
  - The *Notice and Consent* form notifies the New Hire that a background check will occur and the signature gives _____ consent to proceed.
  - The *Release from Liability* must be signed and filled out with date of birth and Social Security number. **Note: It is important that the date of birth and social security number be accepted only after an offer and acceptance of employment.**
  - The *Prescribed Summary of Consumer Rights* form explains the employee's legal rights under the Fair Credit Reporting Act. Under certain circumstances, this form must be given three times: once in the initial offer and acceptance, prior to rescinding any offer as a result of information received during the background check and again if the offer is rescinded at a later date as the result of information received during the background check.

The New Hire must also be notified prior to the date on which the background report is first requested.

Once the background check is completed, Human Resources and the Hiring Manager will be notified. If the background contains adverse information, the Human Resources Department and the Hiring Manager will review the information and make the decision for continued employment or to rescind the offer of employment based in whole or in part on the results of the background check. If the offer for employment is rescinded, the Hiring Manager and Human Resources must:

page 2 of 3

Figure 2.1 Employee background screening policy *(continued on next page)*

- Present the employee with a copy of the background report;
- Give the employee an additional copy of the *Prescribed Summary of Consumer Rights form;* and
- Give the employee a reasonable amount of time (three business days) to dispute the accuracy of the report.

If the employee does not dispute the accuracy of the report, fails to respond within a reasonable time (three business days), or fails to provide feedback to the Human Resource Department and the Hiring Manager within the time period specified, the offer of employment shall then be rescinded. The employee must then be provided with the following:

- Oral, written or electronic notice of the adverse action,
- Name, address and telephone number of the consumer reporting agency that furnished the report,
- Statement that the consumer-reporting agency did not make a decision to take adverse employment action and is unable to explain the specific reasons behind the decision and
- The *Prescribed Summary of Consumer Rights form.*

Any questions or concerns relative to this policy should be submitted to the Director, Corporate Security.

Figure 2.1 Employee background screening policy *(continued)*

infrastructures. The results at electric-generating plants, for instance, could vary from power losses to power surges. For water facilities, the pipelines could be closed to prevent distribution or opened to cause flooding. For sewage plants, the offal could be dispersed into reservoirs that would become polluted. For nuclear facilities, an attack could result in deadly fallout in metropolitan areas.

Lack of Vital Spare Parts

The fifth is a lack of vital spare parts. Core components of our nation's utility/energy infrastructures were built forty or more years ago, yet we do not stockpile the most important large spare parts. If and when key power-generating turbines wear out or are destroyed, for example, it could take six or more months to get replacements built and shipped from overseas, where most of them are manufactured.

TERRORISTS' GOALS

Despite this focus on infrastructures, please don't think that Islamic or other terrorists simply target systems. Instead, they target people, masses of people. They use systems to kill people and, ultimately, to cripple economies. Two tragic examples are the hijacked planes used to attack thousands of people in office buildings during 9/11 and the railroad cars carrying hundreds of people that they blew up in Spain during 2004.

But this strategy of attacking enemies is not new. Islamic extremists began that tradition in the 11th century. The terrorist threat to our nation's utility/energy infrastructures is real. Here's a brief historical perspective.

Jihadism Began in the 11th Century

Let us use Islamic history as an example of the historical development of terrorist activities. According to Elizabeth Cobbs Hoffman, the Dwight Stanford Professor of American Foreign Relations at San Diego State

Figure 2.2 LaCygne power plant

University, martyrdom was an expression of political terrorism as early as the 11th century. At that time a religious leader, an imam called Hassan, established a ruthless cult that became known as the Assassins. He inspired passionate devotion with extreme interpretations of the Koran and a strict requirement of obedience.

To prove their devotion, cult members were required to undertake suicide missions with the goal of murdering the enemies of the imam. "Terrorism was a political act and a sacred religious duty," Professor Hoffman stated. "Angels would carry the broken bodies of the Assassins to Paradise."

The first Islamic jihad, or Holy War, developed in the early life of Islam. The teachings of Mohammad, who died in 632 CE, incorporated both Jewish and Christian teachings. There was one God, Allah, and Muhammad was His Prophet, as was Jesus and the pre-Christian Jewish prophets of the Torah, the Jewish Bible. Both Jesus and the prophets were called "children of the book," and the book was the Koran. It replaced both the Old and New Testaments, spreading in all directions from the Saudi Arabian peninsula.

During this first Islamic jihad, Islam met with little resistance as it firmly established the religion on the Iberian peninsula. Charles Martel of France, the father of Charlemagne, stopped the onward march at the battle of Tours in France. Islam was finally driven out of Spain in 1492 at the battle of Granada, which ended the first jihad.

The Ottoman Turks carried out the second Islamic jihad. That empire caused the downfall of Constantinople as a Christian stronghold and brought an end to Roman hegemony in its various forms. Although the Ottoman Empire was a most successful expansion of Islamic territory, the religion became fractured into warring sects and bitter rivalries. By 1683 the Ottomans had suffered many defeats on land and sea. When they failed to capture Vienna, further territorial ambitions were diminished. Islam shrunk into various sheikhdoms, emir-dominated principalities, and roving tribes of nomads.

By this time, however, a growing anti-Western sentiment became part of the Islamic fundamentalist dogma. Internal failures were blamed on others, particularly Westerners. The new sect that predominated during this new revival was called Wahhabism. Shortly before the beginning of World War I, it came into full bloom under the House of Saud on the Arabian peninsula. This Wahhabi version of Islam has infiltrated the religion itself, now finding adherents in almost all branches and sects, especially the Shiites. What this sect calls for is the complete and total rejection of anything and everything not based in the original teachings of the Prophet.

Wahhabism finds its most glaring practice in the policies of the Afghani Taliban and the Shiite practices of Iran's late Ayatollah Khomeini. Its Field Marshall is Osama bin Laden. He is the leader of the third Islamic jihad. In 1983, this movement got worldwide attention through actions of an Islamic terrorist organization funded by Iran and Syria named Hezbollah. In April of that year, a suicide bomber exploded a truck in front of the American embassy in Beirut, Lebanon. 63 employees were killed and 120 were wounded. In October 1983, 241 U.S. Marines were killed and 81 more were wounded when another Hezbollah suicide bomber blew up an American barracks at Beirut's airport. That December, the American embassy in Kuwait was bombed.

In 1984, the CIA station chief in Beirut, William Buckley, was kidnapped and murdered. During September 1984, an annex to the U.S. Embassy near Beirut was bombed. In June 1985, Hezbollah hijacked an American airliner and made the pilots fly to Beirut, where they held the pilots and passengers for more than two weeks. The jihadists killed an American naval officer who was on the plane and threw his body on the tarmac.

Later that year an Italian cruise ship, the *Achille Lauro,* was hijacked by a group led by Abu Abbas of the Palestine Liberation Organization (PLO). An elderly American passenger who was confined to a wheelchair, Leon Klinghoffer, was thrown overboard.

After several terrorist incidents in Rome, Vienna, and Berlin that were tied to Libya's leader, Muammar Qaddafi, left many Americans dead, the U.S. attacked Libya. Three U.S. citizens who worked in Beirut were killed in retaliation by Palestinian terrorist Abu Nidal.

In December 1988, Pan Am flight 103 blew up over Lockerbie, Scotland, killing more than 270 people, mostly Americans. Two Libyan intelligence officers were tried for planting the bomb; one was convicted, the other acquitted.

I am purposely not focusing on which U.S. President did how much— or how little—after which terrorist action, because I don't want to deflect from a focus on terrorism. I have no desire to even appear to favor Republican or Democratic administrations. Instead, I am trying to provide brief, solely factual summaries of terrorist incidents that occurred during the administrations of Presidents Ronald Reagan, George Bush, Bill Clinton, and George W. Bush.

Terrorism in America

Terrorist action came to the continental U.S. early in 1993. On February 26 a truck bomb exploded in the parking garage of the World Trade Center in New York City. Six people were killed and more than 1,000 were injured. Six Muslim terrorists were convicted and sent to prison.

We subsequently learned that a terrorist Islamic networked named al Qaeda was behind the World Trade Center bombing of 1993, and its leader was Osama bin Laden. During the next two years, Islamic terrorist acts occurred in Israel, Lebanon, Pakistan, Saudi Arabia, Turkey, and Yemen. In Pakistan during 1995, for example, two American diplomats were killed, and in Saudi Arabia that year, five Americans died when a car bomb exploded.

In June 1996, a truck bomb detonated at a building in Saudi Arabia, killing 19 Americans and wounded 240 more. Two years later, American embassies in Nairobi, Kenya, and Dar es Salaam, Tanzania, were bombed, leaving more than 200 dead, including many Americans. Al Qaeda claimed credit for the bombings, both of which occurred on August 7. On October 12, 2000, bin Laden attacked the USS Cole, a destroyer that had been docked in Yemen. The team of suicide bombers killed 17 American sailors and wounded 39 others.

No Stopping Terrorism

This is a brief summary of terrorist actions largely against the United States that led up to the events of September 11, 2001. My purpose in presenting this overview is to show a continuum of jihadist actions. They began in the 11th century, and over the last twenty years, many of them have been directed against American interests.

These attacks by Islamic extremists are continuing. And they will persist. Please don't think otherwise, or you will be helping to put yourself, your family, and our country at risk.

Here is my most important point: We cannot stop terrorism. But we can and we must be alert to the continuing threat, and we can and we must develop a variety of effective ways to make our most important targets far less vulnerable. By hardening our many targets that are at risk, we can make jihadist acts harder to complete, and we can convince some terrorists to try other, softer targets in far-off parts of the world.

There's been much speculation in the media and years of debate in Congress about increased security in response to perceived threats. Here's just one much-discussed example. There is a possibility that terrorists could assemble a dirty atomic bomb, steal a Scud missile, place both on a boat that runs near one of our coastlines, and then shoot the dirty bomb high into the earth's atmosphere. The result could trigger an electromagnetic pulse (EMP) attack.

The pulse generated by the blast would destroy electronics and satellites in its field of vision. A well-executed blast, for example over the Midwest, would not kill people, but it could cause great damage to telecommunications and electronics, including our electric grid, with disastrous consequences. This could happen. My point is not to discount this horrible, possible act, but to focus us in a different direction: The national electric grid and other vital portions of our utility/energy infrastructures can be knocked out much more easily than through an EMP attack.

Similarly, it would be much easier for jihadists to hijack planes that are then crashed into financial centers, government buildings, or sports events where tens of thousands of Americans have gathered. And terrorists would find it easier, far quicker, less costly, and much more likely to succeed if they set their sights on our utility/energy infrastructures. They could accomplish any one of these actions:

- Physically enter one or a number of our inadequately protected chemical, biochemical, electricity-generating, or water-treatment plants and release fluids or pulses that would create havoc over wide areas.
- Dump highly-toxic chemicals into reservoirs that serve any number of the nation's major cities.
- Use the Internet to hack into computer systems that control energy-generation plants, water supplies, oil and natural-gas distribution lines, chemical pipelines, and much more. This cyber warfare can be done from thousands of miles outside the U.S. by people with basic computer knowledge. It's already happening many thousands of times every day as outsiders seek to hack into computers housed at the Department of Defense (DOD) and other military installations. Our military facilities are prepared for this. Most civilian organizations managing our nation's utility/energy infrastructures aren't.

Please note that nothing in this list is news to terrorists. This and all other information in this book is publicly available.

IMPACT COMPARISONS VERSUS OTHER CRISES

We are all still too painfully aware of what happened on 9/11. But please take a moment to think back in time. What if someone had come to you on September 10 and said, "I'm worried that terrorists could hijack some large passenger planes and crash them into skyscrapers and government buildings!"

What would you have said to that person? Beyond what you would have said, more importantly, what would you really have thought about him or her? I think I know your answers. Since 9/11, you and everyone else in the United States have had to adjust his or her thinking. We as a nation not only have begun to think differently; we have also had to change what we plan and how we act.

Now we must do more. We must act more defensively to provide greater protection to our country. So I'm asking you to think about the possibility of a catastrophe in America that makes 9/11 seem as if it's only the first act. Open your mind to the potential misuse of our nation's utility/energy infrastructures by terrorists so that tens of thousands of people are killed and hundreds of billions of dollars are lost.

This thinking can help ensure that the worst-case scenarios won't occur by recognizing the natural and man-made disasters that have happened both here and abroad. The terrorist attacks of September 11, 2001 resulted

Figure 2.3 Transmission substation

in a massive loss of life, destruction of property, and shocks to various economic systems. Although the attack on the World Trade Center ranks as one of the most fatal disasters in U.S. history with 2,976 fatalities, it is half the loss of life that occurred during the Galveston Hurricane of 1900, which took over 6,000, lives or, in another part of the world, the 1991 Bay of Bengal cyclone that killed 139,000. In fact, the 9/11 death toll does not even make the top 100 list of worldwide disasters in the last 100 years.

On a global basis, even more significant in terms of death toll and persistence has been the genocide occurring within Burundi and Rwanda. Approximately 1.2 million people have died in at least six distinct Hutu-Tutsi ethnic exterminations since 1965, most notably 800,000 in Rwanda in 1994.

Terrorism is more of a clandestine process designed to allow small organizations or networks with limited resources to wage war against more powerful governments and nations over long periods of time. As such, attacks range from small to large in magnitude and usually occur in an unpredictable, almost random fashion. The negative economic impact of such a nefarious process is long-term, sustained by not only the actual acts of destruction but also the pure fear of the when, where, and how of that next attack.

IMPACT ON THE UNITED STATES ECONOMY

It's certainly important to study the historical effects of massive outages. But how can we use this information to understand the short- and long-term economic and social impacts if we lose the entire grid for an extended period of time?

When we think about massive outages from a conventional point of view, there is a tendency to point to the following types of widespread yet still localized examples:

- 1965, 1977, and 2003 blackouts in New York City and neighboring areas, with cost estimates of the most recent event ranging up to $10 billion
- August 10, 1996: 7.5 million customers across 11 western states and two Canadian provinces lose power, some for several hours. The estimated economic losses are $2 billion
- August 13, 1999: Downtown Chicago blackout, which shut down about 3,000 businesses in the Loop and caused $100 million in estimated economic losses

- March 18, 2000: 550,000 customers in New Mexico lose power due to grass fire, rendering a key transmission line to be inoperable.

A joint undertaking of Mirifex, a regional business-technology consulting firm, the Center for Regional Economic Issues at Case Western Reserve University's Weatherhead School of Management, and CrainTech measured the economic impact of the 2003 blackout. The event continues to have far-reaching, long-term implications for businesses in the affected region.

Study findings incorporate the input from 129 executive-level managers in Ohio, New York, Pennsylvania, Michigan, Wisconsin, and southern Canada. The key findings are:

- 11 percent of firms say that the blackout will affect their decision-making with regard to either growth or relocation.

- As a consequence of the blackout, over one-third of businesses surveyed (38 percent) said they would be somewhat or very likely to invest in alternate energy systems.

- More than one-third of firms surveyed (34 percent) have no risk management or disaster recovery plan in place.

- Nearly half (46 percent) of the businesses surveyed will invest more in risk management, business continuity, and/or disaster recovery in the future.

- More than a third of the businesses surveyed (35 percent) felt that it was somewhat or very likely that the region's image would suffer as a result of the blackout.

- Two-thirds of the businesses surveyed (66 percent) lost at least a full business day due to the blackout.

- One quarter of the businesses surveyed (24 percent) lost more than $50,000 per hour of downtime—meaning at least $400,000 for an 8-hour day. And 4 percent of businesses lost more than $1 million for each hour of downtime.

- Nearly half the businesses surveyed (46 percent) said that lost employee productivity was the largest contributor to losses suffered due to the blackout.

- Production/manufacturing and customer sales/service were the areas of business hardest hit by the blackout.

How vulnerable is the grid itself? Reka Albert, assistant professor of physics at Pennsylvania State University, has studied the topology of the grid structure and concluded that, although the grid has been designed to withstand random losses of generators and/or substations, its overall integrity may depend on a few key elements. "Our analysis indicates that major disruption can result from loss of as few as two percent of the grid's substations," says Albert, whose research team includes Istvan Albert, research associate in the Bioinformatics Consulting Center at Penn State, and Gary L. Nakarado at the National Renewable Energy Laboratory. One implication of the research is that the identification of a few strategic points in the grid system can bring down the system.

The study, titled "Structural Vulnerability of the North American Power Grid," was published in a recent issue of the journal *Physical Review E*. The researchers constructed a model of the entire transmission grid with over 14,000 "nodes" which include generators, transmission substations, and distribution substations, and more than 19,000 "edges," corresponding to the high-voltage transmission lines that carry power between the nodes. They measured the importance of each substation node based on its "load," or the number of shortest paths between other nodes that pass through it. "While 40 percent of the nodes had a load below one thousand, the analysis identified 1 percent of the nodes—approximately 140—that has a load higher than one million," Albert says.

This high degree of connectivity in the grid system allows power to be transmitted over long distances, but it also allows local disturbances to propagate across the grid. "There are systems to protect the nodes from overload, such as a controlled shutdown to take a substation out if it over-loads or to shut off a generator. In general, these systems do a good job of protecting the nodes," says Reka Albert. "What this model really looks at is the effect of losing a number of nodes in a short period."

If the nodes are removed randomly, the effect on the system is roughly proportional to the number of generators or substations removed. However, the grid quickly becomes disconnected when the high-load transmission substations are selectively removed from the system—if the nodes that have the highest load are removed first, followed progressively by the nodes with successively lower loads. According to the model, a loss of only 4 percent of the 10,287 transmission substations results in a 60 percent loss of connectivity. During a cascading failure, in which the high-load substations fail in sequence, the model shows that the loss of only 2 percent of the nodes causes a catastrophic *failure of the entire system.*

An attack on poorly protected elements of substation automation systems could achieve disastrous effects for the overall grid. In fact, over one-half of the electric utility personnel who responded to a recent Electric Power Research Institute (EPRI) survey believed that a cyber intruder in the information and control systems at an electric utility could cause serious impact on their regions and beyond for more than 24 hours. Easily available resources such as Federal Energy Regulatory Commission (FERC) filings, electricity industry publications, transmission and distribution maps, and the Internet provide ample information on the most critical transmission lines and substations in the power grid. Relatively simple hacking techniques could then be used to locate ports to these points that trigger outages.

Let's turn from attacks on the nation's electric grid to the possibility of a terrorist attack on a nuclear power plant. In 2004, Edwin S. Lyman, Ph.D., of the *Union of Concerned Scientists* conducted a study entitled "Chernobyl on the Hudson? The Health and Economic Impacts of a Terrorist Attack at the Indian Point Nuclear Plant." It reported the following quoted findings:

- The current emergency-planning basis for Indian Point provides insufficient protection for the public within the 10-mile emergency-planning zone in the event of a successful terrorist attack. Even in the case of a complete evacuation, up to 44,000 early fatalities are possible.

- The radiological exposure of the population and corresponding long-term health consequences of a successful terrorist attack at Indian Point could be extremely severe, even for individuals well outside the 10-mile emergency planning zone. Lyman calculated that over 500,000 latent cancer fatalities could occur under certain meteorological conditions. A well-developed emergency plan for these individuals, including comprehensive distribution of potassium iodide throughout the entire area at risk, could significantly mitigate some of the health impacts if promptly and effectively carried out. However, even in the case of 100 percent evacuation within the 10-mile EPZ and 100 percent sheltering between 10 and 25 miles, the consequences could be catastrophic for residents of New York City and the entire metropolitan area.

- The economic impact and disruption for New York City residents resulting from a terrorist attack on Indian Point could be immense, involving damage from hundreds of billions to trillions of dollars and the permanent displacement of millions of individuals. This would dwarf the impact of the 9/11 attacks. The economic damage within 100 miles would exceed $1.1 trillion for the 95th percentile case, and could be as great as $2.1 trillion for the worst case evaluated, based on Environmental Protection Agency guidance for population relocation and cleanup. Millions of people would require permanent relocation.

- The potential harm from a successful terrorist attack at Indian Point is significant even when only the mean results are considered and is astonishing when the results for 95th and 99.5th meteorological conditions are considered. Given the immense public-policy implications, a public dialogue should immediately be initiated to identify the protective measures desired by the entire affected population to prevent such an attack or effectively mitigate its consequences should prevention fail. As this study makes abundantly clear, this population extends far beyond the 10-mile zone that is the focus of emergency-planning efforts today.

- To better understand the amount of loss this nation might sustain, experts conduct modeling efforts. A terrorist attack can lead to significant economic loss, both as a direct consequence of the attack as well as through the secondary or ripple effects felt around the affected region.

Costs of Recovery

A White House report, *The National Strategy for Physical Protection of Critical Infrastructures and Key Assets,* calls for efforts to "develop economic models of near- and long-term effects of terrorist attacks." The report highlights the temporal and cross-sector complexities of modeling such economic damages because the ripple effects across sectors and geographical areas may be significant although difficult to predict. ICF Consulting, in collaboration with Regional Economic Models, Inc. (REMI), explores methodologies and models to measure such potential economic damages.

As part of this effort, the report modeled economic changes resulting from hypothetical terrorist-attack scenarios. It analyzed the impact of a coordinated attack on the electricity- transmission grid in California, for example, resulting in a sizeable loss in electricity supplied to the state. The modeling focused on the economic damages of these attacks, avoiding issues related to measuring the human toll. ICF Consulting examined the scenarios to estimate direct costs on affected industries and calculated the ripple effects on other industries and the economy as a whole. This study revealed the magnitude of the losses and helped determine the best strategies for policymakers to prepare for and mitigate damages.

Another way to estimate the impact on our economy is to simulate an attack. One simulation focuses on a California transmission grid. Based on ICF Consulting's expertise in the energy-transmission and homeland-security sectors, it was hypothesized that the simulated attack on the electric grid caused a severe disruption in the power supplied to California and led to a 25 percent reduction in the availability of electricity. This initial loss resulted in significant economic damage but for a very short period of time.

During the day of the simulated attack, authorities were able to reoptimize the grid and restore power to certain prioritized sectors. This incomplete restoration led to rolling blackouts for an extended period of time. The gradual ramp-up was assumed to continue for two weeks, after which the system was back to its pre-attack state. The economic damage from this attack was calculated for this two-week period. Significant damage resulted from this terrorist attack, leading to a total direct cost of approximately $11 billion.

This initial direct damage predicted by this model had a ripple effect and led to another $7 billion in secondary impact on the state's economy. The total loss of about $18 billion was approximately 1.3 percent of the gross state product of California. The significant negative shocks of this model led to more than 122,000 lost jobs in that state.

To help understand the economic costs of a terrorist attack in the future, we can more closely examine the direct and indirect costs and the timing, both immediate and long-lasting, from the 9/11 attack.

Direct Impacts

Various accounts of the aftermath of 9/11 list well-known statistics, such as the facts that lower Manhattan lost about 30 percent of its office space and that a significant number of businesses ceased to exist. About 200,000 jobs disappeared or fled New York City. Destroyed tangible assets amounted to $14 billion for private enterprises, $1.5 billion for state and local government concerns, and $700 million for federal agencies. Total immediate cleanup and other recovery costs have been estimated in the range of $11 billion. These figures add up to a staggering total direct cost of $27.2 billion.

Indirect Impacts

The security-conscious culture immediately instilled as a direct reaction to 9/11 has had a significant impact on several key industries, including insurance, airlines, tourism and travel, and defense. Let's take a look at each of these industries.

Insurance

Losses from 9/11 have been estimated at between $30 and $58 billion. By comparison, Hurricane Andrew's 1992 ravage of the state of Florida caused about $26.5 billion in losses. The federal government has slated $11.6 billion to help Florida and other eastern states rebuild from the cumulative effects of the four hurricanes that occurred in 2004.

Most insurers increased terrorist insurance premiums at least 30 percent almost immediately after 9/11, and others just dropped that type of coverage completely. These increases have negatively impacted many industries, such as airlines, transportation, construction, tourism, and energy.

Airlines

The airline industry was hit particularly hard. Routes were curtailed or eliminated due to lower passenger traffic, higher fuel costs, and the hassle factor of flying in general. For many short routes, it's just as fast to drive when taking into account increased airport-security measures. The U.S. airline sector lost 40 percent of its value when the market reopened after 9/11 and has never fully recovered. Other related industries that have also been negatively impacted include tourism, automobile rentals, and commercial-aircraft manufacturers.

Defense Spending

The Congressional Budget Office (CBO) recently estimated the cost of occupying Iraq and other operations associated with the global war on terrorism for fiscal year 2005. Assuming current levels of effort, CBO projects that the Department of Defense (DOD) will likely require $55 billion to $60 billion in a new budget. This estimate includes only the costs above amounts budgeted for routine military operations. It does not include any of the cost for reconstruction activities carried out by DOD or other governmental agencies. CBO has estimated the 10-year (2005-2014) cost of the occupation of Iraq and other operations (Noble Eagle, Enduring Freedom, classified activities, coalition support, and activities that cannot be allocated to specific operations) to be between $182 billion and $393 billion.

Some of the immediate economic impacts resulting from 9/11 have been lessened to varying extents, but there are permanent losses that will never be recovered. For example, the diminution of terrorism-related insurance coverage will stymie investment of capital in businesses more vulnerable to terrorist acts due to higher potential risks. Another example is the additional deficits due to accelerating defense needs imposed on the federal budget, which can hamper growth and productivity.

Longer Term Impacts

The ever-present threats of potential terrorism can be analyzed by looking at the components making up productivity—output divided by input. Let's take a look at how these components are affected:

- Most industries will have increased operations and maintenance expenses due to higher spending levels for physical and cyber-security needs, insurance requirements, and corporate taxes. Many high-tech product-development costs will be larger due to increased security features.
- Many firms hold larger amounts of inventory due to increased risk of shipping, including air, rail, and trucking. For example, immediately after 9/11, industries that depended on supplies from foreign countries learned a lesson because they had production interrupted by security-related delays.
- In addition to higher debt-financing costs due to increased business risk, most companies have experienced demand for higher returns from their stock to attract needed equity investment. This in turn has raised the cost of equity capital for business-expansion purposes.
- In order to combat terrorism head on, governmental research and development (R&D) resources have been shifted away from civilian applications to military use. As a result, more of a financial burden is being placed on private enterprises to make up the difference.
- Firms tend to invest closer to home under the potential threat of terrorism, mainly due to increased transaction costs—higher insurance premiums, shipment delays caused by increased security checks, and fear of unstable political situations, reducing the ability of businesses to expand to global markets.
- If and when there is a terrorist attack with long-term consequences, a recent study reported that 40% of Fortune 1000 companies will not survive two years after the attack, unless they are properly prepared.

According to a report prepared by IFC Consulting in Fairfax, Virginia, the cost of the 2003 blackout was between $7 and $10 billion for the national economy "and would been significantly higher had it been caused by a terrorist attack." This study points out that beyond grid malfunctions, there is a "hangover effect... the most significant burden was borne by the tourism industry as people became nervous and avoided travel."

If we talk about the costs of recovery proactively instead of reactively, we change the subject to ways that utilities get reimbursed for security improvements. New legislation and regulatory proceedings in some states have been developed to achieve cost recovery, often involving rate case proceedings. In Michigan, for example, Jeff Pillon is the Chairman of the Energy Data and Security Committee of the National Association of State Energy Officials (NASEO). "Utilities in Michigan are not reluctant to take upfront actions that improve the security of their critical infrastructures, then petition the state for cost recovery," he says. "Commissions here are predisposed to approve these investments. Some utilities take the defensive perimeter approach; others are more aggressive in their pursuit of security by, for instance, building in redundancy that leads to greater reliability. The latter group anticipates all-hazards protection in terms of terrorists who could blow it up or hurricanes that could blow it down."

Michigan is one of the leading states. A July 2004 report prepared for the National Association of Regulatory Utility Commissioners (NARUC) Ad Hoc Committee on Critical Infrastructure concludes, however, that "Most states have not received security-specific cost-recovery requests."

California, Florida, Iowa, Michigan, New Jersey, New York, and Oklahoma are the examples for other states to follow. Chapter 3 presents important initiatives by these states, the federal government, the utility/energy industry, and private enterprise that help to protect the utility industry.

Chapter 3

WHAT IS BEING DONE TO PROTECT THE UTILITY INDUSTRY?

Let's turn this discussion to more positive issues and consider what we are doing to help secure our nation's utility industry.

FEDERAL RESPONSES

Since 9/11, the Federal government has developed some security standards for the utility/energy industry to deter terrorist attacks. These standards primarily focus on cyber security, nuclear security, and government utility facilities that have federal government oversight, such as the Tennessee Valley Authority (TVA), which answers to the U.S. Congress.

Nuclear Plant Security

The nuclear-energy industry is one of the few whose security program is regulated by the federal government. Nuclear plants must meet all federal security requirements, which are determined and monitored by the U.S. Nuclear Regulatory Commission (NRC). The NRC has on-site inspectors at each plant to ensure compliance. At issue is protecting the public from the possibility of exposure to radioactive releases caused by sabotage. By gathering and analyzing intelligence information about incidents around the world, the NRC helps to ensure that nuclear-energy plant-protection regulations are updated in line with potential threats.

The NRC's "design basis threat" defines the threats against which nuclear plants must provide protection. The NRC determines the design-basis threat by reviewing technical studies and information received from intelligence experts and federal law-enforcement agencies The NRC's security regulations ensure that the industry's security force is able to protect facilities against a ground-based threat. The threat is characterized as a suicidal, well-trained paramilitary force, armed with automatic weapons and explosives, that is determined to force itself into a nuclear power plant to commit radiological sabotage by using a biological agent/agents introduced into the facility. Such an action might have the assistance of a plant employee, who could pass along information and help the attackers. But because of the nuclear energy industry's security programs and the defense-in-depth safety strategy, the FBI classifies nuclear power plants as difficult targets due to the strict federal standards that have been put in place post-9/11.

Federal Anti-Cyber-Attack Programs

"The vulnerability of our control systems to cyber attacks has never been more critical than it is today," says Joseph Weiss, KEMA Consulting Cyber Security Practice Leader and a well-known industry expert on the electronic security of control systems. "Control systems and electronic protective relays are essential to the functioning of the infrastructure and key economic sectors," he explains. "While these systems have been designed for performance, minimal consideration has been given for their electronic security. In fact, there have already been more than 30 cases of intentional or unintentional cyber impacts on control systems. Such cyber security breaches can result in damaging and costly safety, regulatory, and business-operation consequences."

For many years, there has been conjecture that terrorists might try to launch attacks on the nation's electricity grids, including hacking into utilities' SCADA (Supervisory Control and Data Acquisition) systems or other core information-technology (IT) or operational infrastructures. To

Figure 3.1 Computer hacker

repel such attacks, new software designed to greatly enhance the cyber security for the utility industry is being tested at a federal-government laboratory in Idaho. The U.S. Department of Energy's Idaho National Engineering and Environmental Laboratory (INEEL) has technology that mirrors real-world utility infrastructures, such as systems, wireless technology and processes.

Several vendors of high-technology equipment provide customers with a single view of systems, along with alarms and suggested processes to avert downtime and potential security problems, supporting an emergency-management system for the utility industry. It enables people at a central source to monitor the beginning of an attack at the router or firewall level. Ron Ross heads a partnership between the National Security Agency and the National Institute of Standards and Technology. He has said that, after 9/11, air-traffic controllers brought down every commercial plane in the air. "If there had been a cyber attack at the same time that prevented them from doing that," he said, "the magnitude of the event could have been much greater." He adds, "A cyber attack can be launched with fairly limited resources."

To verify this point, U.S. intelligence agencies have upgraded their warnings about al Qaeda's use of cyberspace. A National Intelligence Estimate on the threat to U.S. information systems also gave prominence to China, Russia, and other nations. Counter-terrorism analysts have known for years that al Qaeda prepares for attacks with elaborate packages of photographs and notes. A computer seized at an al Qaeda office in Afghanistan contained models of a dam, made with structural architecture and engineering software that enabled the planners to simulate its catastrophic failure.

A terrorist can breach a dam without using tons of explosives. Instead, he or she can use cyberspace.

In 1998, a 12-year-old hacker broke into the computer system that runs Arizona's Roosevelt Dam. Federal authorities said that he had complete command of the SCADA system, controlling the dam's massive floodgates. The 12-year old hacker did this just for the fun of it. This dam holds back as much as 1.5 million acre-feet of water, or 489 trillion gallons. It could wash over a floodplain encompassing the cities of Mesa and Tempe, which have a total population of approximately one million people.

Teams of mock intruders from the Energy Department's four national laboratories devised eight different SCADA attacks on the nation's electrical power grid. All of them worked. Joseph M. Weiss of KEMA Consulting, an expert in control system security, said, "The attackers were able to assemble a detailed map of each system, and they intercepted and changed SCADA commands without detection. In most cases, they were not using anything that a hacker couldn't have access to." There are an estimated 3 million SCADA systems in use throughout the nation. In many of the following chapters, you'll see that, from a security standpoint, these SCADA systems are highly vulnerable. So much so that some utilities are considering building parallel private networks for themselves.

In describing the threat to all types of companies, Richard A. Clarke, President Bush's former cybersecurity adviser, has said, "It doesn't matter whether it's al Qaeda or a nation-state or the teenage kid up the street. Who does the damage to you is far less important than the fact that damage can be done. You've got to focus on your vulnerability. . . and not wait for the FBI to tell you that al Qaeda has you in its sights."

STATE RESPONSES

Some state regulators are waiting for the federal government to set security standards for the industry. But the federal government is not acting. So other states are taking the initiative to develop security standards on their own. The approaches of state governments vary, as they use different homeland-security structures and processes:

- The Pennsylvania House of Representatives charged the Pennsylvania Public Utility Commission and the Pennsylvania Emergency

Management Agency to develop a comprehensive evaluation of the protection of the state's critical infrastructure and to recommend improvements.

- Iowa constructed a Homeland Security Critical Asset Assessment Model to reflect critical infrastructure assets. These include a geographic-information-system (GIS) database of asset criticality and vulnerability.
- Hawaii's Homeland Security Critical Infrastructure Protection approach, led by the State Civil Defense Agency, tapped the state's Energy, Resources, and Technology Division to work with the private/public Energy Council. The latter, comprised of all of Hawaii's major energy companies, certain specialized military units, and various agencies from all levels of government, has prioritized infrastructure facilities according to criticality to national and economic security. Facility vulnerability assessments have been conducted. Guidelines for infrastructure protection and force deployment plans have been developed and a statewide critical infrastructure-protection plan put into place for implementation according to state-developed color-coded threat levels.
- The Maine Adjutant General chartered nine ad hoc homeland-security teams consisting of government and private-sector interests to develop plans and also to identify critical infrastructure needs. In addition, the Maine Public Utilities Commission has signed a confidentiality agreement with the FBI as a secure InfraGard member, and a number of Maine PUC staff members are being granted security clearances.
- Michigan established a Homeland Security Task Force to coordinate all related actions across a broad multidisciplinary spectrum, including federal, state, local, and private organizations. Staff from the Michigan Public Service Commission chairs the CIP committee and the energy subcommittee and has representatives on the communications and transportation subcommittees.
- New Jersey was one of the first states to create a Domestic Security Act shortly after 9/11. The governor has charged the Domestic Security Task Force to oversee the development of a synchronized and comprehensive program of domestic preparedness designed to coordinate all state initiatives to detect, prepare for, prevent, protect against, respond to, mitigate, and recover from any act of terrorism. The task force is comprised of specific subcommittees, including the

Domestic Security Preparedness Planning Group, the Weapons of Mass Destruction Advisory Committee, the Infrastructure Advisory Committee, and various other support groups.

To help meet the homeland-security challenge, every governor has designated a state homeland-security director to coordinate state efforts, some within existing frameworks, such as law enforcement or defense, and some through special homeland-security teams.

New York's Model Program

The people of New York responded heroically to the terrorist events of 9/11. We as a nation continue to respect the physical nature of terrorism, yet we cannot lose sight of the realities and threats posed by cyber attacks. Nor can we ignore the devastation natural disasters have on our technology infrastructure. So whether we anticipate man-made acts that are intentional or inadvertent, acts of nature, or cyber incidents, attacks on our critical infrastructure can be devastating.

New York State, for example, is developing a comprehensive approach to cyber security that:

- Inventories the assets of the critical infrastructure.
- Identifies the vulnerabilities of and potential threats to these assets through the use of technology.
- Focuses on prevention, detection, response, and recovery.
- Is building stronger relationships between the public and private sectors, as well as between the civil and law-enforcement sectors.

The New York State Office of Cyber Security and Critical Infrastructure Coordination (CSCIC) was established in 2002 to address the state's cyber readiness and resilience. CSCIC is responsible for leading and coordinating the state's efforts regarding cyber readiness and resilience; leading and coordinating geographic-information technologies, especially in emergencies, where CSCIC is the single point-of-contact; coordinating the process by which critical infrastructure data is collected and maintained; expanding the capabilities of the cyber incident-response team; and monitoring networks for malicious cyber activities.

Other State Organizations

The National Governors Association (NGA) promotes teamwork as an essential role and responsibility of all state agencies. The NGA suggests that all state government departments' personnel and resources should be viewed as components of the state's overall emergency-management system. The NGA recommends that governors identify a lead agency for each critical infrastructure and charge that agency with developing long-term protection plans, warning notification systems, response action plans, and strategies to ensure restoration of essential services.

Each state has also identified a state Emergency Management Agency (EMA) to coordinate statewide efforts that mitigate, prepare for, respond to, and recover from emergencies. This is an all-hazards approach to any natural or man-made threat to the state. The state EMA manages the state's emergency response and coordination with federal, county, and local officials through the state's Emergency Operations Center (EOC).

UTILITY/ENERGY INDUSTRY RESPONSE

Private industry is doing significant security work with the utility/energy industry, especially in the areas of contact security service and security consulting. Yet gas and electric utilities are under attack for lax security that leaves their control systems vulnerable to cyber and physical threats.

Comments industry consultant Joseph Weiss, "Our systems have been designed to be open, so they don't have the needed level of security. The entire industrial infrastructure was developed for efficiency and production, not security issues." For example, consider the SCADA systems used to manage and operate facilities at electric and gas utilities by opening and closing valves and switches to regulate the flow of energy. Reports from the General Accounting Officer (GAO) and others have identified the threat of cyber attacks against utility control systems, as well as the constantly growing number of viruses and other Internet threats. The GAO has also cited as problems the adoption of standardized technologies with known vulnerabilities and the increased connectivity of control systems to other systems.

Advances in computer security aren't always incorporated into the process-control systems used by the utilities. According to Joe Weiss, "Security vendors such as RSA, Baltimore Technology, and HP know security very well, but if they try to solve control-systems issues, they could cause a lot of outages because they don't understand these systems. And the control-system vendors don't know security."

Individual utilities deploy their own security processes. As a consequence the National Energy Reliability Council (NERC) has drafted a security standard to set uniform security processes. At Ameren

Corporation, a utility in St. Louis, for example, business-technology and operations people work together to set security standards for control systems based on Linux and Windows. The SCADA systems at the Tennessee Valley Authority operate over the utility's own network, and access is limited to plant system engineers.

Electrical Substations

Substations present another major problem when it comes to cyber security for utilities. Substations usually have many hundreds of times more connections than most antivirus software is designed to handle. Yet the real-time operating systems of substations means that they must function without any interruption. The majority of substations were built for a specific task before we had a concern about cyber security. National-security consultants I have talked with know of at least six outages that have occurred because of security breaches. Many of these instances of inadequate security involved components that were connected to the grid.

Figure 3.2 Data center

NERC has recently published new cybersecurity standards to cover thousands of unprotected electrical substations.

State utility regulatory commissions have jurisdiction over the intrastate components of large portions of our nation's critical infrastructures. The following industries contain critical infrastructure:

- Banking and finance
- Electric power
- Emergency services
- Telecommunications
- Oil and gas
- Transportation
- Water

The jurisdictions for these industries vary by state, as they go for nuclear-power plants, wireless telecom providers, enforcement of safe drinking water acts, wastewater and heating utilities, rail and motor carriers, and more. The issues facing these utilities are many and varied, and so is the treatment of different utility sectors by state commissions. Power grids, natural-gas pipeline systems, and telecom networks typically are neither statewide nor nationwide. Most are regional. Generally speaking, state commissions have recognized the regional nature of our critical infrastructures and are therefore cooperating on a regional basis. National Association of Regulatory Utility Commissioners (NARUC) affiliates, such as the Mid-America Regulatory Conference, the New England Conference of Public Utilities Commissioners, and the Western Conference of Public Service Commissioners, are working together.

Natural Gas Facilities

Security of natural-gas utilities has special importance, especially in some states such as California, where more than 90 percent of all new power plants will be fueled by natural gas.

Disruptions caused by loss of natural-gas service could interrupt the flow of many critical functions, including power from these plants, computer use, and other types of communications.

Above-ground natural-gas operations offer obvious targets. But even when pipelines are buried, computerization of energy operations makes them vulnerable to cyber attack. As a result, natural-gas utilities have invested heavily in card access controls, video cameras, remote video transmission, digital video, a variety of alarm systems, photo ID badges, and larger and improved use of security personnel.

To protect themselves from terrorism, here are some of the ways natural gas utilities are being proactive:

- Reviewing physical security surveys of their facilities and improving a wide variety of barriers, guards, communications, lighting, intrusion detection systems, and other measures
- Designating areas that are restricted so that only authorized personnel ever have access. These include computer operations
- Placing critical equipment and operations on higher floors and fitting those areas with sophisticated intrusion-detection systems
- Maintaining a comprehensive access roster
- Escorting all visitors
- Keeping rooms where people congregate locked whenever possible. These include janitorial areas, maintenance closets and shops, telephone and electrical equipment rooms, and even restrooms
- Maintaining emergency supplies and communications equipment in special, not-easily-identified safe rooms

Nuclear Utility Security

When it comes to security considerations, nuclear utilities are a special subject. The following discussion focuses on Exelon, which is one of many companies that are leaders in nuclear-plant security.

Exelon has spent a considerable amount of time and money since 9/11 to protect itself from and prepare itself for a terrorist attack. Among many measures, this company has increased its coordination and communication with local, state, and federal emergency-response agencies and industry organizations.

Exelon has developed both disaster-recovery plans and business-continuity plans to support rapid business resumption after a crisis, such as a terrorist attack or other operational emergency. Meanwhile, the company has significantly increased the security of its power plants. They are some of the most secure industrial facilities in the world. Armed security guards constantly patrol Exelon's nuclear plants, with the help of sophisticated electronic surveillance equipment, to protect the plants from external attack. Internally, Exelon has instituted multiple layers of employee background verification and tighter restrictions on access to the plants.

Nuclear-power plants are some of the strongest man-made structures in the world. They contain multiple layers of protection and are designed to withstand natural disasters and nuclear accidents. A recent Electric Power Research Institute (EPRI) study has shown that nuclear-power plants are likely to withstand a full-force impact from a wide-bodied jet that is fully loaded with fuel. Additionally, the critical structures of a plant are relatively small and would be extremely difficult to hit with an airplane.

This and many other companies in the nuclear-power industry are coordinating their security measures with local and state emergency response operations and law authorities to protect nuclear plants. Exelon and other companies are also working closely with the FBI and other federal agencies to enhance emergency preparedness and plant security.

HOW DO WE MOVE FORWARD?

States need to make more progress in developing and implementing security standards. We cannot wait for the federal government to set national standards.

Progress at State and Federal Levels

Officials at all levels of government have identified state commissions and state energy offices as key players in the national homeland-security effort. The Department of Homeland Security (DHS) is still consolidating most of the more than 40 federal entities currently identified as having roles in combating terrorism. Federal efforts to assist state and local governments on homeland-security issues need to be stepped up and partnerships established to address homeland-security concerns.

Perhaps the most valuable contribution these agencies can make to the DHS mission is to coordinate their linkages to and relationships with the utility and energy companies, the potential targets of the evolving threat. In the majority of cases, these relationships have developed over decades of coordination and cooperation to ensure reliable energy service and fuel.

DHS can leverage successful state and industry partnerships and facilitate the transfer of best practices among the states. This process would leverage DHS's expertise and resources and have the potential to create successes in less time and at lower costs.

We need a greater public-private dialogue to define the best approach for particular industrial sectors and types of vulnerabilities. There is distrust on the part of private industry about revealing information to other companies—including competitors—and government at different levels. There is no easy answer to this issue of trust.

But we must keep one thing uppermost in our minds. When our nation's critical infrastructure is attacked, the primary responsibility for response to and recovery from terrorist attacks will fall to cities, counties, states, and private industry. Private companies own almost all the utility/energy infrastructures that are targets for terrorism.

Inducing industry to play its critical role in homeland-security activities—to invest in systems for reducing their vulnerabilities and to develop and manufacture counter-terrorism technologies that may not have robust commercial markets—is likely to require new regulatory requirements, financial incentives, and/or voluntary consensus agreements.

A National Regulatory Research Institute (NRRI) survey of state commissions suggests that virtually all have some authority with respect to security for jurisdictional utilities, and most report greater activity in that area since 9/11. The NRRI survey revealed that most utilities are reluctant to share security-related information with commissions, even though most commissions have the ability to protect such information. Integrating state commissions and state energy offices into emergency planning and response is essential. NRRI has suggested a number of possible state commission actions related to homeland security including:

- Collaborating with federal and other state agencies.
- Facilitating inter-industry communication and information exchange.
- Examining utility security preparedness efforts.
- Providing incentives for security and reliability enhancement.
- Considering adequacy of peak-load management and curtailment processes.
- Reviewing priorities for rebuilding efforts if key infrastructure is disabled.

But how can coordination be assured? It requires effective sharing of threat intelligence, vulnerability assessments, and protective measures in critical areas and regularly exercising capabilities. The federal government has designated the National Infrastructure Protection Center (NIPC), housed in the FBI, as the agency responsible for coordinating threat information nationally through Information Sharing and Analysis Centers (ISACs).

For example, NIPC communicates threat information pertaining to electric-power systems to the U.S. Dept. of Energy and the industry North American Electric Reliability Council (NERC) for dissemination through the NERC-operated Electric Sector ISAC (ES-ISAC). The use of

Figure 3.3 Information technology (IT)

sector-specific information mechanisms results in information "stovepipes" where information distribution is artificially limited. The General Accounting Office (GAO) has found that intelligence sharing "is a problem between the federal government and the states."

The InfraGard program was established by NIPC to provide a cross-sector private-public forum related to physical and cyber threats to critical infrastructure. However, the InfraGard program is not uniformly developed nor promoted. In sectors where rigid information stovepipes exist, state commissions possess the appropriate combination of expertise across different sectors, an understanding of critical infrastructure in local and regional areas, and insights into possible vulnerabilities and interdependencies between critical infrastructures. Some state commissions have begun to focus on improving information flow. The Virginia Governor's Office, for instance, has recommended that the State Corporation Commission participate on a team to further communication and intelligence sharing between government and the utility industry.

Coordination also requires the ability to communicate on a continuous basis. Some state commissions have developed contact names and information for law enforcement, emergency management, utility facilities, and commission personnel and regularly share that information with those partners. Con Edison's experience in New York in September 2001 highlighted the requirement that prior arrangements must be in place for utility personnel who might need prompt access to disaster areas. Only then can they render a scene safe for rescue and other first-responder personnel, even if the area is secured as a crime scene.

Also recognizing the need for reliable sharing of critical information, NARUC's sister organization for state energy offices, the National

Association of State Energy Officials (NASEO), has developed a comprehensive network of Energy Emergency Information Coordinators (EEIC) for each of the states and territories. NASEO has put this information into an electronic listserv that provides an emergency information-broadcasting function. In addition, NARUC and NASEO have established a communications network for key state personnel called the Energy Emergency Assurance Coordinators (EEAC) system. Through this network, state personnel can exchange information and coordinate with each other and the Office of Energy Assurance (OEA) when there are energy emergencies. The EEAC system also enables communication in advance of energy-supply disruptions and emergencies.

The Role of Insurance Companies

The economic consequences of 9/11 include the realization that terrorism losses are potentially too costly to be fully covered by the private sector alone. In this environment, the Terrorism Risk Insurance Act (TRIA) became law in 2002. This law:

- Provides a federal backstop for terrorism risk.
- Requires primary insurers to make terrorism coverage available to commercial policyholders on the same terms, conditions, and limitations as other covered types of loss.
- Has limited the exposure of insurance companies' catastrophic losses from terrorist acts, TRIA has had a stabilizing influence on the economy.

Without TRIA, I think many insurers would reduce—not increase—their exposure to the risk of terrorism. There would therefore be growing gaps in coverage. Currently, for example, the price of terrorism insurance for the energy industry is higher than for any other sector when measured as a percentage of property premiums by industry. Costs would likely increase dramatically without TRIA.

TRIA will expire unless Congress acts to extend it. With this extension, alternative approaches can be more fully tested, including capital infusions, catastrophic terrorism bonds, and risk pooling.

WHO PAYS?

Utility companies are recouping some of the costs of increased security from their customers through higher rates. "From a homeland-security perspective, I think utility customers want to make certain that the critical infrastructures they use are protected," said Jeffery Pillon, a staff member for Michigan's Public Service Commission. "And if the companies, on their own initiative or based on industry guidance, take action to increase security, that gives at least a basis for allowing those expenditures to be included in rates, as long as the costs are reasonable," said Pillon, who is also chairman of a critical infrastructure subcommittee for the National Association of Regulatory Utility Commissioners.

State regulatory bodies oversee and set prices for most utilities and decide whether to grant rate increases in a series of hearings known as "rate cases." The proceedings mirror court cases. Exhibits are introduced and experts are called to support and rebut claims. Transcripts often run into the thousands of pages.

But when all is said and done, it can be hard to determine how much of each customer's gas, electric, and water bills is attributable to security costs. Typically, these costs are only a small part of most rate-increase requests; some of which run into the millions. The fact is that there is no one-size-fits-all security plan for utilities. Even within the same industry, security measures can vary from region to region, depending on topography, likelihood of attack, and other parameters. And what is needed varies too, from installing better padlocks, heavier chains, elaborate motion-detection devices, and complex cyber-security systems to extensive construction projects that harden critical infrastructures.

In Pennsylvania, for example, American Water won a $30 million rate increase that included some incremental security costs attributed to 9/11. The typical monthly residential water bill for a Pennsylvanian is about $38 per month, and less than $1 of that is related to enhanced security costs. Consumers Energy Company in Michigan won approval for higher rates to pay for $24.5 million in security measures at its nuclear plant. The improvements were required under federal guidelines, and nuclear plants across the country are seeking rate increases to cover their enhanced security expenses.

A utility in Iowa won rate increases for $1.6 million in security costs. Iowa-based MidAmerican Energy, for example, serves 460,000 people in the state. It won a gas rate case that included $780,375 in security costs, which added just 27 cents per year to consumers' bills.

A study by the National Regulatory Research Institute, the official research arm of NARUC, found that utilities in 22 states have filed to recover security costs.

Increased rates are only one way that utility/energy companies can be compensated for their greater security costs. Subsequent chapters will discuss additional revenue-generating methods.

Chapter 4

THE POST–9/11 SECURITY ASSESSMENT PROCESS

NEW SECURITY NEEDS

Because of heightened security requirements within the United States and the volatility of the business environment, utilities are seeing the need to move quickly to enhance their security. Utilities are also advising their commercial, industrial, and residential customers that they can depend on their service.

Yet according to a number of government agencies and national security experts the utility/energy industry is vulnerable to breaches of security and ranks high on the list of potential targets by terrorists. Utilities are constantly reminded of how easy it is to access their facilities and control systems. In addition, we have all heard that it is impossible to protect a utility because most have remote facilities and their infrastructure can extend from dozens to hundreds of miles.

We have also heard that an attack on the utility industry or one using utilities to terrorize cities, could have devastating effects on our nation, our citizens, and the economy. It is impossible to completely protect a utility from an attack, be it a cyber attack or a physical assault on a facility or its infrastructure.

The truth is that no one can totally secure a utility. We all know that. However, making a utility tough to penetrate is possible. A number of state regulators, for example, are looking at, and some are requiring, security assessments to be conducted at utility facilities. This must be done before a utility can request a rate adjustment to pay for enhanced security.

When utilities are confronted with pre-employment screening or security checks on vendors, they frequently push back: "We have known these folks for years." In fact, the utility industry has always been very community-oriented. It prides itself on employing only the best. It is not uncommon for a utility employee to have devoted 15 to 20 years of service to one company.

Another concern of some top executives at utility companies is that conducting background checks will involve the company in legal issues. But as long as the background checks stay within the guidelines of the Fair Credit Reporting Act (FCRA), legal issues should not occur.

Let's face the new reality: We in the United States, and in many other countries around the world, are living with cells of al Qaeda operatives, as well as everyday criminals, computer hackers, and angry former employees. We must act in ways that make our vital industries and our country more secure. In the context of this chapter, for example, it's critical to screen all new employees and vendors to ensure that they are who they say they are. Here's what a professional security assessment will address:

- Physical security
- Information and cyber security
- SCADA and DCS systems
- Communications security
- Electric-grid security
- Distribution security
- Generation security
- Biological/chemical issues that include an anthrax assessment

The security-assessment team, which is usually made up of experienced consultants, needs to include recognized experts in security as well as technical experts who have worked in the utility industry. Conducting an assessment is one thing. Arriving at sound solutions and meaningful results is paramount. A professional security-assessment team should be knowledgeable about current technologies and able to recommend financially viable options for implementing solutions.

Many utilities are having security assessments done as a regular course of business, and they are putting recommended security improvements in place. As one consequence of these actions, the majority of utilities are struggling with revenue commitments to pay for improvements and enhancements without compromising their productivity and day-to-day operations. Many utilities look at a range of security alternatives and settle on improving physical and cyber security.

A developing trend is to put in place a *security collaborative* where several utilities share in the cost of an assessment. This approach has been used nationally for years by the utility industry, especially in new technology offerings and shared services. By pooling money and services cost-effective upgrades to technology and ways to deliver services can be achieved. Among the actions this cost-effective approach enables is to have security assessments conducted on a number of facilities sequentially. As recent news stories confirm, in order for a security plan to be as successful as possible, it is critical to *create a security culture* at all levels of the utility, from the CEO on down. This can be accomplished through on-the-job-training seminars conducted by a security professional in conjunction with law-enforcement personnel The more eyes and ears the utility has, coupled with employees trained in what to look for, the better the chances of having a successful security program.

The simple truth is the utility/energy sector operates like a Swiss watch. Damage one small component and the watch will cease to function. Destroy a major refinery, severely damage the natural-gas delivery system, sink some tankers, successfully conduct a cyber attack on telecommunications and electricity delivery grids, sicken the people who keep the system functioning, and then attack. Those are the scenarios that can cause the finely-tuned system to cascade into catastrophic collapse.

Understanding the threat and knowing your vulnerabilities are only part of the answer to these troubling issues. Planning for the worst-case scenario and building that plan takes many skills. Those skills exist, and now it's vitally important to devote the necessary time.

PHYSICAL SECURITY ASSESSMENT PROCESS

Everyone in the security business knows about the three Gs: guns, guards, and gates. But they no longer buy the security this industry and our country need. Physical security assessment of utility/energy facilities consists of many components and can be tasking because of facility locations. It is very common, for example, for many utility facilities to be located in the most remote rural areas. The good news is that most utility/energy companies have communications capabilities in place that enable monitoring at some level at even the most remote facilities.

A thorough physical security assessment will look at hundreds of areas so that all assets can be monitored from a central location on a 24/7 basis. I know consultants who use physical-security audits that include up to 400 review questions that can lead to successful monitoring of all the most important assets. Here are only a small percentage of the review questions in a comprehensive physical security audit:

- What is the total number of employees?
- How many are full-time, part-time, temporary, or contract employees?
- Are background investigations conducted prior to employment?
- Are current photographs maintained of all employees?
- Do you have a card-access system installed?
- Must everyone who wants to access the main entrance to the facility use an identity card with his or her photograph plus a special PIN number?
- In the case of enhanced security, must those who want access also provide a unique biometric characteristic, such as a fingerprint, hand scan, or iris pattern?
- How soon after termination are on employee's access rights removed from card access and from computer access?
- How many shifts are there?
- Do you use outside maintenance crews?
- Do you inspect all packages entering and leaving the building?
- Is government work done at this facility?
- Is the facility subject to natural disasters—hurricanes, tornadoes, floods?

- What types of fences, walls, buildings, and other physical perimeter barriers are in place?
- Is there an adequate clear zone along both sides of fencing?
- Can vehicles drive up to fencing?
- Are perimeter barriers regularly maintained and inspected?
- How is the facility alarmed, and which employees have access codes?
- Is the facility alarmed for smoke, fire, penetration, motion, heat, glass breakage, and other factors?
- How many exterior doors are there on the premises?
- Are they all alarmed?
- Is CCTV used on all exterior doors?
- Is CCTV used to observe the exterior perimeter?
- Is sensitive material stored inside administrative offices?
- If so, how is that material secured?
- What security procedures are followed in warehouses?
- How are truck drivers controlled?
- How are visitors monitored through the facility? Are they escorted? Do they wear badges? Are they required to sign in and out?
- What types of locks are used, and how often are they changed?
- Do you have a disaster-recovery plan?
- Do you have a business-continuity plan?

INFORMATION AND COMPUTER SECURITY ASSESSMENT PROCESS

The utility/energy industry has excellent technologies in place at their corporate facilities, power plants, and many rural facilities. Most corporate facilities, for instance, have installed redundancies that include backup technologies, duplicate systems in the event that one fails. Generally speaking, it all works well in terms of serving commercial and industrial clients and residential customers.

But there are two major downsides: The majority of these systems are not running on secure networks. And they are easily hacked. As mentioned in prior chapters, if the computer systems of these facilities are accessed and damaged electronically, utility service to major areas of the United States can be shut down. These issues need to be addressed immediately.

This sense of urgency was recently highlighted when Patrick H. Wood III, Chairman of the Federal Energy Regulatory Commission (FERC), held a private meeting with leading electric-company officials. Mr. Wood is reported to have told them that they need to focus more heavily on cyber security. In public appearances with diverse audiences, he has raised this pressing issue. The former chief of counterterrorism in the Clinton and Bush administrations, Richard A. Clarke, has also tried to alert the nation to these dangers. Mr. Clarke has said, "A sophisticated hacker or group of hackers...could probably get into each of the three U.S. North American power networks and could probably bring sections down..."

I personally know security consultants who have successfully hacked into the computer systems of utility companies. And I know security experts have hacked into one or more of the three North American power networks. It has been done by people on our side. I am concerned that it will be done by people who are against us.

Existing systems can be enhanced to prevent hacking without replacing the entire current technology. Consultants who conduct information-security audits can address issues after asking more than three hundred questions about information and computer security. The following lists ask a small portion of the most important questions. The first list asks questions to obtain overall information security at a central site:

- Has a formal disaster-recovery plan been developed, tested, and updated annually?
- Have operating procedures been established to maintain business if the computer center is not operational?
- Have emergency procedures been developed for bomb threats, fire, high winds, and flooding?
- In the building where the main computers and information are housed, are the walls, roof, and floors constructed of noncombustible materials to reduce the chance of fire?
- Are the walls near critical equipment and media constructed of materials that cannot be easily penetrated?
- Are ceilings watertight to prevent leaks from floors above?
- Has an uninterruptible power supply (UPS) and/or a backup generator been installed? If so, is it or are they tested quarterly?

- Do backup systems have adequate resources to enable emergency work to run?
- Does the computer center, especially the CPU, have its own separate power supply and power cables?
- Are emergency power-shutdown controls installed at all fire exits?
- Are lock-and-key control procedures for computer and information storage areas reviewed every six months?
- Are sign-on and sign-off procedures elaborate and airtight, so that it's highly unlikely that an outsider can tap a line and take over access with full privileges?
- Are there systematic procedures for the handling and disposing of sensitive materials?
- Does the room layout allow supervisors to watch the progress of classified jobs at all times?
- Are combinations to safes, locks, and padlocks changed at least annually and whenever key personnel leave?
- When an employee is terminated, is a checklist used to assure that all important materials are returned before the final paycheck is issued?

As you know, utility/energy substations and extensions of infrastructures are very often located in remote sites. Additional issues need to be addressed to help make them more secure. The following questions related to overall information security at remote sites:

- Has the physical protection at each remote terminal location been reviewed and evaluated, whether or not these locations are under direct main-office supervision?
- Are the programs, data files, and operating software of standalone computers backed up and stored off premises?
- Is access to sensitive stored information—whether it's electronic or printed— safeguarded programs, and confidential files tested a number of ways to deter unauthorized access? For example, can the computer system be entered from a remote location outside of regular business hours but using a legitimate name, password, and location?
- Are sign-on and sign-off procedures as elaborate and secure at remote facilities as at the central site?

COMPUTER SECURITY

Recently, I talked with a top executive of a utility in one of our largest cities about the security of his company's vast computer system. He said that they were far advanced in that area. When I asked him to be specific, he responded, "Well, we have a very good email system."

As you probably know, email is a major gateway to viruses, worms, and computer hackers. Instant messaging is even less secure and must be discouraged in all sensitive areas. Yet there are hundreds of ways to increase the security of email, software, and other types of computer applications. The following questions address some of the most important ways to increase security for email, software, and other applications:

- Are users trained on computer-security awareness and procedures? For example, are all employees that use computers trained in the basics of handling spam? If not, attachments from unknown senders can introduce viruses that can spread in milliseconds throughout the entire enterprise.
- Do employees know that they should never write passwords on notepaper that they attach to their computer and that they should never disclose passwords?
- Is virus-protection software updated frequently by every employee or through the system network?
- Have standards been established and implemented to address hardware, software, operations, documentation, and security?
- Have standards been established and implemented to provide for audit trails, program approvals, and activity logs for all programs dealing with sensitive data?
- Is it possible to alter data without producing an audit trail?
- Are audit logs routinely reviewed for unusual activities?
- Has access-control software been installed to protect the data on the hard disk from unauthorized access?
- Have responsibilities been reviewed to ensure that no one person is totally responsible for any single activity dealing with sensitive information?
- Have firewalls been deployed in ways that maximize security? If there's any question, rather than relying on a false sense of security, hire a security consulting firm to run tests on the firewalls and access points.

- If your utility/energy company isn't large enough to have 24 /7 intrusion detection, should you consider outsourcing this service to a computer security company?

BIOCHEMICAL AND ANTHRAX SECURITY ASSESSMENT PROCESS

Utility/energy companies have made very little if any effort to secure their facilities from a chemical attack. These attacks can come through the mailroom or, as mentioned in prior chapters, introduced into a power plant to transform the facility into a weapon of mass destruction. In order to prevent this from happening, chemical sensors need to be installed in all areas of a utility, from the corporate headquarters to remote facilities. There are technologies that can sense the presence of agents, and these technologies can contain them to prevent a massive exposure. Most of these technologies are new and are being tested.

PRE-EMPLOYMENT AND ANNUAL SCREENING

The utility/energy industry has an excellent track record of hiring good people. But the post-9/11 environment requires a closer look at who is hired and a verification that applicants are, in truth, who they say they are. Identity theft—stealing the identity of another through access to credit cards, Social Security number, or other documents—is increasing significantly, and terrorists also have access to fake identification credentials that allow them to pass most basic background checks.

Utilities that are downsizing because of deregulation are contracting out more and more services. As a result, contractors and other service vendors have increased access to facilities and infrastructures, yet they are seldom subject to background checks.

A further issue concerns utilities that do background checks yet limit them to pre-employment screenings that are not updated on a yearly basis. Just because an employee passes a pre-employment screening does not mean that he or she has been cleared for the full term of their employment.

The following is a minimal list of areas that should be screened for new hires and contractors. It is important to review or re-screen employees and contractors on an annual basis:

- Social Security verification and validation. This means that the federal government issued Social Security cards and that the users of that card are who they say they are.
- Credit check, especially if it relates to the new hire's position
- State sexual-offender check
- National criminal search
- State criminal search
- County courthouse search
- Federal courthouse search
- Motor-vehicle check, especially if it relates to the new hire's position
- Education verification
- Employment verification, including reference checks
- Credential verification—this pertains to professional licenses and other claims of accredited expertise
- Drug testing

I recommend that background checks be conducted by a reputable service provider that uses court runners. This is the best way to ensure that on-site verification at courthouses will be done. A number of consultants to the utility/energy industry use only databases. When this method is used, there is a good chance that convictions will escape notice, especially in rural areas.

DUE DILIGENCE SCREENING

I have already mentioned conducting background checks on contractors. The same should apply to other service providers in the form of due diligence screening. Just because their employees check out does not mean that the service provider as a company is worthy of your trust. When a utility/energy company enters into a long-term service contract with a provider, a due-diligence check will validate if the company is sound from a business perspective, and it will also ensure that it is not in any way affiliated with a terrorist group or doing business with terrorist operatives. My experience has shown that some companies do business with terrorist-affiliated groups yet are ignorant about these relationships. This due diligence also prevents a utility/energy company from taking a PR bath for doing business with a terrorist-affiliated company.

Ness Group International – Dallas, TX
WWW.NESSGROUP.COM
LNESS@NESSGROUP.COM
214.415.9687 Office

NOTICE

This form, which you should read carefully, has been provided to you because EMPLOYER either may request, has decided to request, or may request during the term of your employment, consumer reports or investigative consumer reports in connection with your application for employment or during the course of your employment (if any), with EMPLOYER. Such reports may contain information concerning your work habits, job performance and criminal history (if any). The types of reports that may be requested, include, but are not limited to, criminal records checks, court records checks, driving records, and / or summaries of educational and employment records and histories. The information may be obtained from public record sources, current or former employers, or other sources. EMPLOYER will use the information contained in such reports solely for employment related purposes and will not use the information contained in the report in violation of any applicable federal or state equal employment opportunity law or regulation.

If EMPLOYER requests an investigative consumer report, you will receive additional notice and information regarding further rights pertaining to investigative consumer reports.

Before EMPLOYER takes any adverse action on your application or employment, based in whole or in part on the information contained in a consumer report, EMPLOYER will give you a copy of the report and a summary description of your rights in connection with such reports.

CONSENT

I have carefully read and understand this notice and consent form and, by my signature below, authorize EMPLOYER and/or its agents to obtain a consumer credit report or investigative consumer report in connection with my application for employment. I further understand that this consent will apply during the course of my employment with EMPLOYER, should I obtain employment, and that such consent will remain in effect until revoked in a written document signed by me.
I further understand that any and all information contained in my application or otherwise disclosed to EMPLOYER by me may be used in obtaining the consumer reports or investigative consumer reports

page 1 of 2

Figure 4.1 Notice and consent *(continued on next page)* requested by EMPLOYER and confirm that all such information is true and correct.

I hereby release EMPLOYER, including its officers, employees, and agents, from any and all liability for damages arising from collecting consumer reports regarding me for employment purposes.

Dated _____

Signature _____

Printed Name _____

Social Security Number _____

DOB _____

If you wish to revoke this consent at any time during the course of your employment (if any) with EMPLOYER, please contact the Human Resources Department.   page 2 of 2

Figure 4.1 Notice and consent *(continued)*

SECURITY CULTURE CHANGE TRAINING

Before 9/11, it was often the practice to have very basic security programs in place in the larger utility/energy companies. At that time, however, they were considered overhead by management and, in most cases, they were poorly funded. In addition, their visibility, compared to other departments and divisions, was very limited. Post-9/11, security programs have seen some changes, and a few have been elevated so that they are managed by vice presidents with security services that are well utilized.

In order to make this necessary transition and avoid the "big brother syndrome" - of constant security review, in many cases management must change the security culture of the company. This starts with training at all levels of the organization. The oldest story in the security world is told about the CEO and other members of senior management who are not the models for a successful security program. When that's the case, the staff will not follow their lead, and the security efforts will not be successful.

On the other side of the coin, if you have a medium- to large-size utility/energy company, it just makes good sense to have everyone looking

A Summary of Your Rights
Under the Fair Credit Reporting Act

The federal Fair Credit Reporting Act (FCRA) is designed to promote accuracy, fairness, and privacy of information in the files of every "consumer reporting agency" (CRA). Most CRAs are credit bureaus that gather and sell information about you—such as if you pay your bills on time or have filed bankruptcy—to creditors, employers, landlords, and other businesses. You can find the complete text of the FCRA, 15 U.S.C. 1681-1681u, at the Federal Trade Commission's web site (http://www.ftc.gov). The FCRA gives you specific rights, as outlined below. You may have additional rights under state law. You may contact a state or local consumer protection agency or a state attorney general to learn those rights.

- **You must be told if information in your file has been used against you.** Anyone who uses information from a CRA to take action against you— such as denying an application for credit, insurance, or employment— must tell you, and give you the name, address, and phone number of the CRA that provided the consumer report.

- **You can find out what is in your file.** At your request, a CRA must give you the information in your file, and a list of everyone who has requested it recently. There is no charge for the report if a person has taken action against you because of information supplied by the CRA, if you request the report within 60 days of receiving notice of the action. You also are entitled to one free report every twelve months upon request if you certify that (1) you are unemployed and plan to seek employment within 60 days, (2) you are on welfare, or (3) your report is inaccurate due to fraud. Otherwise, a CRA may charge you up to eight dollars.

- **You can dispute inaccurate information with the CRA.** If you tell a CRA that your file contains inaccurate information, the CRA must investigate the items (usually within 30 days) by presenting to its information source all relevant evidence you submit, unless your dispute is frivolous. The source must review your evidence and report its findings to the CRA. (The source also must advise national CRAs—to which it has provided the data—of any error.) The CRA must give you a written report of the investigation, and a copy of your report if the investigation results in any change. If the CRA's investigation does not resolve the dispute, you may add a brief statement to your file. The CRA must normally include a summary of your statement

page 1 of 3

Figure 4.2 Fair Credit Reporting Act (FCRA) form *(continued on next page)*

in future reports. If an item is deleted or a dispute statement is filed, you may ask that anyone who has recently received your report be notified of the change.

- **Inaccurate information must be corrected or deleted.** A CRA must remove or correct inaccurate or unverified information from its files, usually within 30 days after you dispute it. **However, the CRA is not required to remove accurate data from your file unless it is outdated (as described below) or cannot be verified.** If your dispute results in any change to your report, the CRA cannot reinsert into your file a disputed item unless the information source verifies its accuracy and completeness. In addition, the CRA must give you a written notice telling you it has reinserted the item. The notice must include the name, address and phone number of the information source.

- **You can dispute inaccurate items with the source of the information.** If you tell anyone -- such as a creditor who reports to a CRA—that you dispute an item, they may not then report the information to a CRA without including a notice of your dispute. In addition, once you've notified the source of the error in writing, it may not continue to report the information if it is, in fact, an error.

- **Outdated information may not be reported.** In most cases, a CRA may not report negative information that is more than seven years old; ten years for bankruptcies.

- **Access to your file is limited.** A CRA may provide information about you only to people with a need recognized by the FCRA—usually to consider an application with a creditor, insurer, employer, landlord, or other business.

- **Your consent is required for reports that are provided to employers, or reports that contain medical information.** A CRA may not give out information about you to your employer, or prospective employer, without your written consent. A CRA may not report medical information about you to creditors, insurers, or employers without your permission.

- **You may choose to exclude your name from CRA lists for unsolicited credit and insurance offers.** Creditors and insurers may use file information as the basis for sending you unsolicited offers of credit or insurance. Such offers must include a toll-free phone number for you to call if you want your name and address removed from future lists. If you call, you must be kept off the lists for two years. If you request, complete, and return the CRA form provided for this purpose, you must be taken off the lists indefinitely.

- **You may seek damages from violators**. If a CRA, a user or (in some cases) a provider of CRA data, violates the FCRA, you may sue them in state or federal court.

page 2 of 3

Figure 4.2 Fair Credit Reporting Act (FCRA) form *(continued on next page)*

The FCRA gives several different federal agencies authority to enforce the FCRA:

FOR QUESTIONS OR CONCERNS REGARDING:	PLEASE CONTACT:
CRAs, creditors and others not listed below	Federal Trade Commission Consumer Response Center–FCRA Washington, DC 20580 202-326-3761
National banks, federal branches/ agencies of foreign banks (word iNational " or initials "N.A." appear in or after bank's name)	Office of the Comptroller of the Currency Compliance Management, Mail Stop 6-6 Washington, DC 20219 800-613-6743
Federal Reserve System member banks (except national banks, and federal branches/agencies of foreign banks)	Federal Reserve Board Division of Consumer & Community Affairs Washington, DC 20551 202-452-3693
Savings associations and federally chartered savings banks (word "Federal" or initials "F.S.B." appear in federal institution's name)	Office of Thrift Supervision Consumer Programs Washington, DC 20552 800-842-6929
Federal credit unions (words "Federal Credit Union" appear in institution's name)	National Credit Union Administration 1775 Duke Street Alexandria, VA 22314 703-518-6360
State-chartered banks that are not members of the Federal Reserve System	Federal Deposit Insurance Corporation Division of Compliance & Consumer Affairs Washington, DC 20429 800-934-FDIC
Air, surface, or rail common carriers regulated by former Civil Aeronautics Board or Interstate Commerce Commission	Department of Transportation Office of Financial Management Washington, DC 20590 202-366-1306
Activities subject to the Packers and Stockyards Act, 1921	Department of Agriculture Office of Deputy Administrator- GIPSA Washington, DC 20250 202-720-7051

page 3 of 3

Figure 4.2 Fair Credit Reporting Act (FCRA) form *(continued)*

for potential security breaches. In a way that is similar to safety programs, encourage every employee to report security violations. It is essential that those who report violations not fall into the "whistle blower" category for doing so.

Ness Group International
Professional Loss/Risk Consulting

Ness Group International – Dallas, TX
WWW.NESSGROUP.COM
LNESS@NESSGROUP.COM
214.415.9687 Office

RELEASE FROM LIABILITY

I certify that the information given by me to _____ is true and complete to the best of my knowledge. I understand that, if I am employed, discovery that I gave false information may result in immediate dismissal.

I hereby authorize any person, educational institution, government entity or company I have listed as a reference on my employment application to disclose in good faith any information they may have regarding my qualifications and fitness for employment.

I hereby release _____, including its officers, employees, directors and agents from any / all liability for damages arising from collecting consumer reports regarding me for employment purposes.

Dated: _____

Signature: _____

Printed Name: _____

NOTE: Do not fill in Social Security Number and Date of Birth until an offer of employment has been given and accepted.

Social Security Number: _____

Date of Birth: _____

_____ **recognizes the confidentiality of information gathered during the background screening process and will not release any information to persons not directly involved in the hiring process.**

Figure 4.3 Release from liability

Creating a security-culture training program can accomplish this improvement if it is mandatory for new employees and yearly attendance for refresher courses is required for all employees. The best way to create a security culture that lasts is to follow the advice of experts who have successfully put these programs into practice. Doug McQueen is one of them. The Director of Leadership Development at American Electric Power in Columbus, Ohio, Doug has been involved in culture change at three major utilities.

"The best way to change culture is to do it under the radar," Doug advises. "Start by ingraining messages into established programs. In leadership classes, for example, have discussions around such questions as: If we operated in a very secure environment, what would it look like? What would it feel like? Employees will respond to this type of presentation."

Doug also says that employees like to have a say in decision-making, and they are always interested to know what's in it for them, whatever the program. So you can start with what's in it for the group, then go to what's in it for the employee? One way to accomplish this is to repeat messages over and over. Here's one important message: "None of us is as smart as all of us." Stronger decisions come from groups, not just individuals. Managers need to keep repeating this message when talking with their teams.

"You can use the security scenario to teach other skills, such as leadership," Doug continues. If an employee sees or hears that there's a violation of security practices, for instance, he or she will show leadership by treating these infractions as violations of safety practices. That, in fact, is what they are. "It's in the individual's interest and the group's interest to correct the unsafe practice," Doug states.

To improve the security culture, attack it on multiple fronts, including leaders, managers, team activities, and training programs. Then celebrate successes around it. When areas are made safer and more secure, for example, feature them in utility publications. Catch people doing well, and then reward them with letters of commendation and photographs on the utility website. These are among the ways to reinforce positive behavior.

Of course, it will take some time to change the security culture at utility/energy facilities. Throughout this process, the objective is to educate everyone at all levels of the organization on the following:

- Security policies, procedures and guidelines
- Business-continuity planning and disaster-recovery planning (discussed in Chapter 5)

- Interaction with federal, state, and local law-enforcement and emergency-service personnel
- Intelligence briefings
- Business ethics
- Public and media relations
- Confidentiality and classified material and documents
- Management philosophy and support for security
- Responding to emergency situations and training of staff as first responders

Although the lists of security culture-change training objectives will vary depending on the organization, the goal is to create a culture in which security becomes a critical part of all management and staff job descriptions. This new strategy contrasts starkly with the traditional approach of looking at security as a threat or as overhead that nobody really wants. Once in place, this vital new strategy helps to create a safer, more secure environment for all. And it encourages and rewards management and staff for their expanded efforts to make our utility/energy facilities and infrastructures more secure.

Chapter 5

SECURITY CRISIS MANAGEMENT PLANNING

As utilities move into this new era of security, the adoption of sound policies, procedures, and guidelines is of the utmost importance, along with an updated crisis-management plan to address new scenarios that could possibly occur. Development of crisis-management, business-continuity, and disaster-recovery plans needs to include reference to federal, state, and local law-enforcement and emergency-services personnel to ensure that all bases are covered before, during, and after an incident.

Yet more than three years after 9/11, nearly one-quarter of all types of American companies are operating without developing, testing, and implementing contingency plans. Surprisingly, about 25 percent of the companies surveyed that are based in New York City and Washington, DC, are unprepared. South Florida has the lowest number of unprepared companies, 15 percent, while Los Angeles has the worst record, for 30 percent of companies there are working without contingency plans.

HOW TO INCREASE SAFETY

Let's determine some of the best ways to increase our safety. We start by turning to a highly experienced clinical psychologist, Dr. Dick Harper, who lives in southern Utah and works with companies in the western United States. Dick often recites the following quotation by Louis Pasteur: "Chance favors the prepared mind." With that statement, Dr. Harper advises that if we're better prepared, we will respond more effectively to whatever happens.

Our tasks, Dr. Harper also suggests, are really more than preparing security risk assessments and crisis-management plans. They are a fundamental part of what we must do to accomplish the bigger picture: Be more secure, increase our safety, and maintain our well-being. In that spirit, I ask you to think of risk assessment, business-continuity planning, crisis-management planning, and disaster recovery as positive actions to take that will help keep your utility/energy company, your community, and our country safer.

SECURITY RISK ASSESSMENT PROCESS

Let's take a wide view of risks so that personnel in your utility/energy facility and its infrastructures can be well prepared, no matter what happens. Not only for terrorism, but also for fire, natural disasters—high winds, flooding, lightning, earthquakes, tornadoes—computer hacking, and other malicious acts, including burglary and civil unrest.

Robert Huber is Founder and President of USG-Inc., based in Minneapolis, Minnesota, one of the nation's largest business-continuity-planning companies. According to Bob, "Rather than worry about a particular event, instead develop plans that prepare you and your company for a variety of major events. You'll sleep better at night knowing that you're ready for a call that might come at 3:00 AM."

Begin by assessing your geographical location to determine what types of risks your company is subjected to. Fire is a possibility everywhere, for example, but flooding or earthquakes are more likely in some areas.

Materials and Procedures to Have in Advance

There are certain procedures and products that you should have on hand to be prepared for most events. Consider the following the basics for key executives and all employees:

- Top executives and support staff need the phone numbers and email addresses of key employees and customers.
- A key executive should have remote access to voice mail and a special number on which he or she can record emergency messages. All employees should know to call this number in an emergency.
- Similarly, that executive should use remote access to change the company greeting and to forward calls to a remote site if necessary.
- Employees should know whom to contact in case they see an emergency or asuspicious activity.
- Employees should also have family crisis plans, including child care in case of emergency.

Top executives and employees at many levels will need access to some basic products. Some employees will need to follow system procedures. Here are some of the most important examples:

- Even if your local building codes don't require them, you need emergency lights that turn on if the power goes out.
- Similarly, your computers need backup systems, with tapes stored at a different location, as well as surge protectors and battery backup systems.
- A National Oceanographic and Atmospheric Administration (NOAA) weather radio with a special tone can alert security personnel to severe weather forecasts.
- Keep in stock whatever supplies and inventory you need for business continuity.
- Check your insurance policies for different types of coverage, including earthquake, flood damage, other natural disasters, and terrorism.
- Keep emergency supplies on hand and distributed throughout the facility, including flashlights with extra batteries, first-aid kits, tools necessary to open locked or blocked exits, and food and water for a reasonable period of confinement.

Outside Risks

Outside your walls, risks from others can include crime, weather and natural disasters, fire, civil unrest, and terrorism. To minimize these risks, make sure that your facility has on hand:

- Fences and gates that prevent intrusion

- Security cameras that monitor entrances, exits, parking structures, any roof access, stairwells, hallways, and all other access points, including loading docks
- Security-card access to all entrances
- Security guards in the main lobby, viewing TV monitors, and walking throughout the facility
- Exterior lighting that illuminates all usable areas at night

Inside Risks

You'll also need to review your buildings and infrastructures. Some procedures and products include the following:

- Be wary of suspicious letters and packages, including those that have no return address; have odors, stains, or anything protruding; or are unusual in weight, size, or shape.
- Security cameras need to monitor people and freight elevators, stairwells, hallways, and all other entry points.
- Backup generators need to be available in case of power failures.
- You'll need to have readily available blueprints of your facility's power- distribution system.
- The data-center room should have a raised floor, an environment that is separately controlled from the main building's system, and a power-distribution system with its own Uninterruptible Power Supplies (UPS) system. The battery gels of the UPS system should be filled.

CRISIS MANAGEMENT PLANNING

Security risk assessment naturally leads to crisis-management planning, business-continuity planning, and disaster recovery. I will discuss these subjects separately in this chapter because I think the information will be presented more clearly that way. Yet all of these subjects are closely intertwined as part of security risk-management issues.

"The most effective business continuity plans (BCPs) or recovery plans are designed and developed for the company for use by responders who are well trained," says Michael D. Ness, Director of Corporate Security and Risk Management for GiftCertificates.com, a Seattle-based provider of gift certificates to businesses and consumers. He continues, "For these plans to be effective, they must be rehearsed on a periodic basis. Unfortunately, most organizations don't do this."

To encourage your company to develop a plan that will minimize your facility's and its infrastructures' exposure to natural and man-made disasters, here's Mike's advice. "Consider the plan an insurance policy. If something happens at one location, you now will have another place to go. You'll have an alternate location, ways to bring your servers up, and plans to contact all employees, industrial and commercial clients, and consumer customers. Take a worst-case scenario, develop plans that act as an insurance policy, and you won't be nervous about getting a call at in the middle of the night."

Crisis Management Team

To develop or improve an existing crisis-management plan, establish a team that becomes responsible for maintaining the good name of your company. In addition to security personnel, this team should have some members from top and middle management, as well as at least one person from the human-resources department and from marketing and/or public relations.

The crisis management team will be in charge of:

- Ensuring the safety of all employees at all facilities.
- Handling emergencies of all types that occur during business hours. In this action they will coordinate activities with appropriate law-enforcement and emergency-services personnel.
- Creating and releasing appropriate information to clients, media, and employees.
- Dealing with legal issues that may arise out of an emergency, turning matters over to company attorneys if and when necessary.
- Counseling employees after an unexpected event.

The crisis-management team is best assisted by a special communications team. In advance of an event, you'll need to see that a call-response team is created.

Call Response Team

This communications team must to stay in close touch with the crisis-management team, so members should be made up of carefully selected managers, marketing employees, and other people who find it easy to communicate under pressure. They must to be schooled in security issues and policies so that they are more likely to use good judgment when giving information to employees, the public, and the media.

The call response team receives:

- Information that has been prepared by a company spokesperson and distributes it.
- Calls from company employees who request updates on progress.
- Calls from clients and customers who request updates on progress. These calls may be rerouted to the appropriate business-unit teams if they are prepared to take them.
- All other calls from outside.

REVIEW OF EMERGENCY PREPAREDNESS

Once you have established a security risk-management process and have put a crisis-management plan in place, I suggest that you conduct a review of this process and plan. The review will help troubleshoot your emergency preparedness.

Michael Steinle, Director of Security Consulting Services at Clarence M. Kelley and Associates, Inc., Kansas City, Missouri, has reviewed dozens of emergency-preparedness plans using strategic assessment tools. Michael says, "There are three major steps to take in emergency preparedness and vulnerability assessment. They are resource analysis, hazard/target analysis, and gap analysis."

Resource Analysis

As you survey your resources, you need to focus on analyzing your human resources, your documents, and your physical assets.

Human Resources
Among the most important human resources, determine if your team members include the following; if they don't you must add people with important experience to the team:

- Arson and bomb experts
- Representatives from local police, fire, and community emergency-response teams
- Specialists for hazardous materials

Document Review

There are several documents to be reviewed, including:

- Contingency plans
- Emergency operations plans
- Existing mitigation plans
- Radiological/chemical stockpile/hazardous-material emergency plans
- Statewide domestic-preparedness strategy

Physical Assets

The physical assets include critical facilities, as well as different types of sites and systems in your geographic area. To be fully prepared, the appropriate personnel at your facility will need to be familiar with a great many assets that can help in an emergency, including:

- Computer backup facilities
- Corporate centers for relocation
- Electricity production-, transmission-, and distribution-system components
- Emergency medical centers and hospitals
- Emergency operations centers and emergency medical services (EMS) facilities
- Law-enforcement agencies
- Local, state, and federal government offices
- Military installations, including National Guard and Reserve
- Oil and gas storage and shipment facilities
- Radiological-materials disposal sources
- Red Cross emergency-response sources
- Telecommunications information
- Transportation infrastructure components, including airports, interstate highway information, oil and gas pipelines, railheads and rail yards, bus terminals, and tunnel and bridge information
- Water-distribution systems

Hazard Analysis

Potential hazards include those caused by terrorists or technological accidents. Those caused by terrorists can include:

- Agriterrorism
- Biological/chemical agents and release of hazardous materials
- Conventional explosives/arson/incendiary attacks
- Cyber terrorism
- Nuclear bomb and radiological agents

There are a large number of technological hazards that should be considered. The most important ones include:

- Industrial accidents at facilities
- Industrial accidents as part of transportation systems
- Problems with the Supervisory Control/Data Acquisition (SCADA) system
- Problems with other components of the critical infrastructure

There are many ways to categorize potential hazards. Here is a methodology to better prepare for them by addressing their different characteristics, such as:

- Application mode, which include the human acts or unintended events that cause the hazard
- Duration, the length of time the hazard occurs at the target
- Whether the hazard is dynamic or static, which indicates whether the event will expand, contract, or remain confined in time, magnitude, and space. The effects of a cloud of chlorine gas escaping from a storage tank, for example, are subject to the velocity and shifts in direction of the wind
- Mitigating conditions that can reduce the hazard's effects. Examples include earth berms, which provide protection from most bombs and sunlight, which dissipates some biological agents.

Vulnerability Assessment

To gauge the vulnerability of your utility/energy facility, let's use Dick Guilmette's Risk of Loss Quadrant, first introduced in Chapter 1.

Gap Analysis

With a solid understanding of your resources and assets, and now of your vulnerabilities, too, it's time to address the potential gaps in your preparedness.

BUSINESS CONTINUITY PLANNING

Whatever might happen to your utility/energy facility and its infrastructure, you will want to do everything possible to keep it running. Business-continuity planning (BCP) is designed to do that. This overview of different types of plans is designed to have actions in place and well rehearsed. So, as Bob Huber says, "If you get a call at 3:00 AM about trouble, you'll know what to do, and you'll know that others are ready to do what's necessary."

Let's begin by agreeing on definitions about the different levels of disasters. Remember that we're not concerned about the cause—fire, dangerous weather conditions, problems with current or former employees, or terrorism. Instead, we're focused on increasing the security of this vital component of our nation's critical infrastructures.

Levels of Disaster

Look at different types of disasters according to how long they last. By using that perspective, there are three levels of disaster:

- Level 1 disaster—the problem can be corrected within 24 hours.
- Level 2 disaster—the problem will take up to 48 hours to correct, but then the situation should be back to normal.
- Level 3 disaster—it's likely that the problem will take more than 48 hours to rectify. Understand that national statistics show that if, for example, a building is unusable for more than 48 hours, employees are likely to be out of the structure for 11 weeks.

Key Business Continuity Tasks

A number of key personnel and procedures will be involved in your business-continuity plan. Begin by focusing on key people:

- Identify essential personnel.
- What departments are they are in?
- What assets (information, intellectual property, documents) are they responsible for?
- What is the day and evening contact information?
- How are they to safeguard and, if necessary, transmit their assets during or after a major event?

Now turn your attention to your utility/energy facility and its infrastructures:

- Identify the critical business functions of your facility and its infrastructures.
- What activities must be maintained if at all possible?
- What actions must be taken to maintain them?
- Who will take those actions?
- What resources will those people need?
- How will they access those resources?

In some cases, you'll need to revert to Plan B. To identify those cases and what to do about them, you must analyze the impact of different events on your business.

Conduct a Business Impact Analysis

Here are some key questions to answer:

- What if one of your essential people is missing from work for an extended period?
- What if one of your critical business functions is out of commission?
- What will be lost without this person and his or her function?
- What's the damage if this critical business function is offline for an extended period?

Answer those questions and you'll determine how much effort needs to be put into creating a recovery process. To help respond to the people questions, determine how specialized that person is and whether or not others within the company can quickly cover for him or her:

- Can you elevate an employee or combine job functions so that more than one person performs the missing person's responsibilities?
- Do you need to go outside the organization to bring in a consultant?
- Should you outsource to a specialist?

Track Revenue Flow through Business Units

Different types of events can interrupt the flow of information throughout your organization. One vital type of information that it will be necessary to track is revenue flow, which is normally available through information about clients, payroll, products, and services. How will this information be accessed in case of a disruptive event?

No one person can answer all these questions and address these many issues. Your company will need to form a variety of teams that can prepare responses in advance of an event and, with practice, put the plans in place during and after a disaster to get the utility running again ASAP.

Create Recovery Teams

The type, number, and composition of recovery teams will vary depending on the nature and consequences of an event, as well as the disruption it can cause to your utility/energy company and its infrastructures. Here is a list of the most important teams, many of which are discussed in more detail within this chapter:

- The information-technology (IT) team, which handles computers and all digital data, including its storage
- Business-unit teams, for both revenue-producing and non-revenue-producing functions
- Crisis-management team
- Call-response team
- Logistics team

A large part of forming these teams is identifying their areas of responsibility, then getting more specific by determining the tasks each team must perform. You'll find initial information on these subjects under the respective teams. What's really important to recognize here is that recovery-team training exercises must be planned and practiced. You can't simply write about what-if scenarios and responses to them. Your utility must act upon the plans and test them to determine what must be improved. People need to become comfortable with the responsibilities they might well need to perform under great pressure.

After practicing the actual exercises, the company needs to produce a recovery document that can, in fact, help the organization achieve business continuity when faced with a disruptive event. But don't file it away in a drawer or computer folder. Distribute it widely to members of all teams and update it as often as necessary.

Only by following all these steps can you know that, whenever you might get a call about a disaster at the office, you and every team member will know what to do.

DISASTER RECOVERY PLAN

What would you and other managers do if one of your main buildings was consumed by fire? What would happen if the phones were down for many hours, the computers were hacked, and valuable records were destroyed?

In order to prepare for these emergencies, you need to create teams that will respond efficiently and effectively. These disaster-recovery teams include:

- IT
- Business units
- Crisis management
- Logistics
- Damage assessment and salvage
- Vendors
- Call response

Right after a disaster occurs, these recovery teams must take a number of important actions. Here are some of the most vital actions of some teams. The IT team must:

- Get the necessary equipment to recover essential IT services, business-critical data, and software.
- Activate all appropriate recovery contracts. (The logistics team will make the phone calls.)
- Install the equipment.
- Load the operating systems and the required application software.
- Set up the LAN and the WAN.
- Test the system.
- Set up PC workstations for business units including LAN.
- Modify applications as necessary to synchronize correctly with data

The business-unit teams must:

- Assist with setting up workstations.
- Contact clients to make sure that they understand that the company has had a disaster and are working with temporary limitations.
- Begin organizing individual workloads so that as much as possible can be carried out without computers for a period of up to 72 hours.

The logistics team needs to:

- Notify the recovery site that a disaster has been declared and prepare for the arrival of recovery workers.
- Arrange for offices and lodging, if necessary, in a nearby hotel or conference facility.
- Assist in hiring personnel with appropriate skills to assist in carrying out the recovery process.
- Receive requests for supplies, food, and equipment from other teams, place those orders with the right vendors, then see that the delivery is sent to the appropriate recovery site.
- Contact company vendors as needed.
- Assist IT as needed, for example, by calling hardware vendors or consultants as well as the IT recovery site.
- Act as the company shipping and receiving department.
- Provide information about damage estimates from the disaster site.
- Oversee attempts to recover salvageable items from inside the damaged facility.
- Arrange transportation for salvageable material, company personnel, contract personnel, as well as purchased goods and materials.
- The leader of the logistics team monitors the recovery of critical documents, tracks progress in obtaining items the company has stored at the offsite location, and keeps track of tasks that are time-critical such as transfers of cash or files at particular times of day.

Summary of Emergency Procedures

When a disaster has occurred and the utility is operating under emergency conditions, the following steps should be followed according to the level of damage:

- Level 1 (problem): It will take up to 24 hours to resolve this event, which involves minor equipment breakdown, partial loss of network, major program error, contaminated databases, and similar problems.
- Level 2 (emergency): Resolution of this type of event will take between 24 and 48 hours. There has been moderate damage to the facility, the infrastructures, and/or the computer equipment.
- Level 3 (disaster): It will take more than 48 hours to resolve this event because of extensive damage to a major facility and/or computer

equipment. All functions and personnel are moved to one or more recovery sites.

Level 1 Responsibilities

Here are the actions that should take place for a Level 1 event:

- An employee immediately notifies his or her supervisor about a problem.
- The supervisor confirms the problem, determines what actions are appropriate, and estimates the length of time necessary to resolve the matter.
- When the issues have been resolved, the supervisor notifies managers that normal operations may resume.

Business Continuity Coordinator

As you'll soon see, Level 2 and Level 3 events require that someone in a key supervisory position be designated the Business-continuity coordinator. He or she:

- Determines the level of the disaster.
- Decides on the appropriate actions to be taken.
- Contacts members of the recovery management team to activate the business- continuity plan.
- If necessary, provides the time and place to set up a command center.
- After the disaster-recovery business-continuity plans are operational, determines when teams can be disbanded and when employees can return to work.

Level 2 Responsibilities

For this serious event, there are more actions to be taken by more people. Note that one supervisor on each shift should be designated as the business-continuity coordinator. At the start of a Level 2 event:

- An employee should immediately contact his or her supervisor about the problem.

- The supervisor identifies that an emergency has occurred and immediately notifies the business-continuity coordinator.
- This coordinator confirms the level of the event and determines what actions should be taken and by whom. The coordinator also contacts members of the recovery-management team, which activates the business-continuity plan.
- If appropriate, the business-continuity coordinator provides the time and place to set up a command center and generates instructions for other employees and for the use of the recovery site.
- Team leaders call all team members, offering instructions on dress, meeting location, and other information supplied by the business-continuity coordinator.
- Team members report their findings and estimates only to their team leaders, who are their points of contact. Otherwise, different versions of loss estimates and inconsistent recovery information can get to the business-continuity coordinator.
- Operations continue under the temporary emergency conditions until the business-continuity coordinator determines that employees can safely return to the area of the event. At that time, the business-continuity coordinator tells managers that normal operations may resume and all teams may be disbanded.

Level 3 Responsibilities

For Level 3, the most serious disaster, the leader of the response is the business-continuity coordinator. The full disaster-recovery and business-continuity plans are put into action:

- As soon as an employee identifies a problem, he or she notifies a supervisor.
- The supervisor determines that a disaster has occurred and immediately notifies the business-continuity coordinator.
- The business-continuity coordinator confirms that it is a disaster and determines the appropriate level of action. He or she contacts members of the recovery-management team, which activates the business-continuity plan.
- If necessary, the business-continuity coordinator provides the time and place to set up a command center and offers instructions to employees and about the use of the recovery site.

- Team leaders place calls to all team members and provide instructions on dress, meeting location, and any other information from the business-continuity coordinator.

- Team leaders give their team members clear instructions that include:
  - Obtaining damage estimates
  - Estimating salvage probabilities
  - Performing salvage operations
  - Estimating recovery time
  - Setting up operations at the recovery site
  - Resuming operations under emergency conditions

- Team members report their findings and estimates only to their team leaders. This procedure prevents multiple versions of loss estimates or inconsistent recovery information from reaching the business-continuity coordinator.

- Team leaders give regular updates to the business-continuity coordinator.

- After initial disaster-recovery operations are underway and preliminary salvage assessments have been made, the business-continuity coordinator might disband some teams.

- Operations continue under the temporary emergency conditions until the business-continuity coordinator determines that employees can safely return to the area of the event. At that time, the business-continuity coordinator tells managers that normal operations may resume and all teams may be disbanded.

As various areas of your company become operable, either manually or by computer, leaders of different teams and the business-continuity coordinator will monitor and report on the status of all recovery tasks. These reports will continue until the utility/energy company resumes normal daily activities.

According to Bob Huber of USG-Inc, "The average cost per hour of downtime for a Fortune 1000 company is $1 million per hour. It takes 22 to 42 hours, for example, just to reload the servers of this type of company." So it makes excellent sense to develop and practice the plans will get your utility/energy company running again as soon as possible.

SECURITY POLICIES, PROCEDURES- AND GUIDELINES

What's most important in security crisis management planning is to train key employees so that, in a disaster, you and they will respond according to plan. The emotions of the event won't interfere with tackling the problems, resolving them, and getting back to work as soon as possible. The different plans are designed for training and reference. They are backup that should be updated but not relied on during an emergency. When a disaster occurs, it's too late to read about actions to take. That's the time when you need to know what to do as well as when and how to do it right the first time.

Here are some of key policies, procedures, and guidelines that will enable your utility/energy facility and its infrastructures to better deal with a disaster:

- Top executives must determine which information and documents are critical to the company's survival, and which they can afford to lose or can reconstruct.
- These executives must determine where these documents are located within your buildings.
- Pinpoint the most important documents on floor plans that are part of your business-continuity documents.
- Make copies of the key documents and store them offsite in a secure facility.
- Make backups of the most important databases and stores these offsite, too.

I've said it earlier, but it's so important that I'll repeat the point here: For your company to be more secure, these different plans must be tested and improved through practice. What are their strengths and where are their weaknesses? It's best to begin with small tests, then build up to full implementations. These comprehensive tests should be conducted at least twice annually.

And please realize that no plan, no matter how well thought-out and tested, will be implemented carefully if top management doesn't consistently demonstrate that it considers these plans a fundamental priority. Even the very best plans for business continuity, crisis management, and disaster recovery will work only if top management says that they must work.

The CIO Council has developed some guidelines for the most serious incidents, Level 5, which include "Fully Integrated Procedures and Controls[†]." Following are six bullets from this document:

- There is an active enterprise-wide security program that achieves cost-effective security.
- Security vulnerabilities are understood and managed.
- Threats are continually reevaluated and controls adapted to the changing security environment.
- Additional or more cost-effective security alternatives are identified as the needs arise.
- Costs and benefits of security are measured as precisely as practicable.
- Status metrics for the security program are established and met.

Selecting a Disaster Recovery Center

Bob Huber reports that 50 percent of his business-continuity clients recover at hotels that are set up for conferences and conventions. "With quick-ship agreements, you can get computer equipment shipped overnight, and data in storage can be retrieved within hours."
How do you determine which disaster-recovery facility will be best for your company? Here are some criteria for selection:

- The location of the facilities should be close enough to be convenient for your employees but not so close as to be automatically affected by the same event.
- The center should provide the management services your utility/energy company is most likely to need.
- The quality and capacity of those services and the backup systems they have in place during an emergency should be appropriate.
- Conduct due diligence on the financial viability of the center as well as any vendors likely to provide needed services.
- Your access to these needed services during an emergency should be adequate. Will you be able to get all the services you need despite other disaster-recovery customers that are simultaneously being served?

[†] Federal Chief Information Officer's Council, Federal Information Technology Security Assessment Framework, prepared by the National Institute of Standards and Technology (NIST), November, 2000.

- Do you have the flexibility during a disaster to use additional services if needed?

My final words on this subject are the most important: Always protect people, including your employees and members of the surrounding community. Create plans that offer the swiftest and safest exit from your facility and the affected area. That's the most important priority in any and every crisis-management plan.

Chapter 6

CRIMINAL JUSTICE COMMUNITY ROLES

In Chapter 2, I asked you to broaden your thinking to include terrorist attacks that could be much worse than 9/11. The purpose of that request was to help make America stronger and safer by expanding the ways we practice security. In this chapter, I reach back many years to a familiar concept in the language of security professionals but one not yet fully integrated into practice by management in many industries.

One of the great practitioners of security, Bob Johnson, has talked about it for years, and he has called it Securemetrics. Bob has said, "We as a nation must include security as a discipline in products, services, business schools, our thinking, and our way of life. Security needs to be a partner in the business process." One of the leading security consultants in the nation, Ray Humphrey, lives in the Boston area but practices across the country and around the world. He has a similar perspective to Bob Johnson on this vital subject. "We must learn to perceive security not as a cost but as a benefit," Ray states. "In order to do this, security must be costed from the start."

There are a number of ways that executives at utility/energy facilities can improve the security of their plants and infrastructures. They can call upon the great security experience and resources of their colleagues at other utility and energy facilities throughout their state, region, and nation. And within their state, there is tremendous benefit in working on contingency-planning activities and coordinating more closely with law enforcement and other segments of the criminal-justice system.

In this chapter, I'll show ways to transform security from a cost to a benefit, as well as ways to save time and money as you create synergy through greater interaction with other utility/energy facilities and members of the criminal-justice system.

COMMUNICATION AND COORDINATION

The president of Duke Power, Bill Coley, has talked about interaction between utilities and government agencies at many levels. Here's what he has said. "[There is a] need for effective communication and coordination among companies, governments, and the public. While this is a simple concept, the complexities involved in carrying it out make it anything but simple.

"Let's consider the kinds and numbers of agencies involved, for example. At the federal level, [Duke Power is] involved with the Office of Homeland Security, the FBI, the Department of Defense, Federal Emergency Management Agency (FEMA), Federal Energy Regulatory Commission (FERC), and the Nuclear Regulatory Commission (NRC), among others. On the state level, we interface with the governors' offices in North Carolina and South Carolina and state law-enforcement agencies. And then there's local government, including the mayor's office, city and county government, local emergency services, along with local law enforcement.

"In addition to the governmental agencies, there are trade and industrial organizations related to the utility industry, including NERC, SERC, EEI, NEI, and others. And finally there are the electric utilities themselves—including investor-owned, electric cooperatives, and municipal systems. Considering everyone involved, establishing communication and coordination protocols for such a broad group of agencies and companies will be quite a challenge. It is critical for these entities to talk—to communicate effectively—if we are to prevent acts of terrorism."

THE CRIMINAL JUSTICE COMMUNITY

The components of the criminal-justice system include local law-enforcement and emergency personnel and plus state and federal organizations. I'll give special focus to rural communities, too, because utility/energy infrastructures located away from population centers have important and distinctive needs.

It is critical that the utility/energy industry develop a close liaison with the entire criminal-justice community and emergency-services personnel before putting security programs in place at all their facilities. The majority of utility/energy infrastructures, power plants, mines, pipelines, and other facilities are located in rural America. To assume that the criminal-justice community can respond in a timely manner to major incidents in all rural areas is a mistake.

If however, the criminal-justice community is involved in the development and implementation of security programs, proper planning can correct any deficiency before an incident occurs. This may require financial/equipment support from the utility/energy industry including the hiring of additional criminal-justice community staff.

Figure 6.1 Security camera

Examples of support are:

- Providing financial support for addition law-enforcement personnel to ensure 24/7 responses in a timely manner.
- The purchase of modern technologies (such as bomb-detection equipment, biochemical equipment, communications equipment, and computers).
- Availability of patrol vehicles, ambulances, and aircraft.
- Financial support to assist the criminal-justice community in soliciting intelligence as it relates to the utility industry.
- Modern equipment, such as vests, weapons, gas masks, and assistance to ensure adequate advanced training can be obtained as it relates to the utility industry. It is important to note that the utility industry has industry-specific security issues that require specialized training, especially in the areas of chemical exposure, steam systems, fires and explosions, and high-voltage exposure.

As you can see, if a utility were to set up a professional security program, it is not correct to assume that it is going to get the level of support it might expect from federal, state, and local authorities. However, by working closely with the criminal-justice community all these issues and more can easily be resolved. The criminal-justice community needs to understand how the utility industry functions in order to put crisis-management plans in place. Once prepared, handling threats should not be an issue.

DEPARTMENT OF HOMELAND SECURITY

The largest federal agency was established in 2003 when Congress approved the creation of the Department of Homeland Security (DHS) by consolidating all or parts of 22 federal agencies. The new department seeks coordination that secures the nation's critical infrastructure through partnerships with the public and private sectors.

DHS is definitely a step in the right direction and could have a positive impact on the utility/energy industry. It should be noted, however, that most of the department's efforts have been focused on airports, ports, and nuclear plants, as well as correcting intelligence issues on a global basis.

Of particular interest to readers of this book, all of DHS's efforts are totally dependent on the utility/energy industry. If we were to experience an attack on our nation's grid system, the homeland-security programs would not function without electricity.

It is critical, therefore, that DHS personnel work with the utility/energy industry to ensure that their current programs will function if and when another terror attack occurs. The following are some examples of DHS programs and utility/energy industry needs to address as soon as possible:

- DHS, along with federal regulators and the Department of Energy (DOE), needs to reach out to the utility/energy industry to immediately get in place security programs that protect our nation's grid system, especially in the area of cyber security.
- Basic security standards need to be developed for the utility/energy industry, and they require financial and tax support from the federal government.
- Intelligence programs need to be established that ensure that the utility/energy industry not only reports activities but also receives briefings from the DHS.

Since the 9/11 attacks, federal agencies have required vulnerability assessments for most water utilities, natural-gas transmission and distribution systems, and electric transmission systems. Emergency response plans have been modified and updated, and security guidelines have been developed at many utilities.

Accountability for utility operations, including security, is mandated by the Sarbanes-Oxley Act of 2003. Guidelines for utility responses to changing DHS threat levels for physical and cyber security have been developed and updated by the North American Electric Reliability Council (NERC). DHS funding is provided for upgrading security at some utilities, and most utilities will be affected by the National Incident Management System (NIMS) and local Incident Command Systems (ICS).

While the focus of many of these security programs is on preventing acts of terrorism and attacks by other extremists, an effective utility security program must also focus on two other potential threats: disgruntled employees and angry customers. How much damage can a disgruntled employee cause? Consider the results of a cyber-crime study conducted by the U.S. Department of Defense. It indicates that the average cost of an incident caused by an outside hacker is $56,000, while the average cost of responding to a malicious cyber attack by an insider is $2.7 million.

Regardless of how effective your customer-service program is, every utility will have an occasional angry customer. Angry customers have vandalized utility equipment, including vehicles, threatened and assaulted utility employees, and been responsible for other malicious acts, such as bomb threats.

DHS's five levels of terrorist risk are a model for utility and energy facilities to follow:

- Green is low risk. Protective measures should focus on everyday facility assessments and the development, testing, and implementation of emergency plans.

- Blue is guarded risk. At this level, protective measures focus on activating employee and public information plans, exercising communication channels with response teams and local agencies, and reviewing and exercising emergency plans.

- Yellow is elevated risk. Focus on increasing surveillance of critical facilities; coordinating response plans with allied utilities, response teams, and local agencies; and implementing emergency plans as appropriate.

- Orange is high risk. Limit facility access to essential staff and contractors, and coordinate security efforts with local law-enforcement officials as appropriate.

- Red is severe risk. Consider closing specific facilities and redirecting staff resources to critical operations.

For more information about these threat levels, as well as news and information from the Department of Homeland Security, please visit www.dhs.gov.

One of the keys to an effective utility-industry response to a terrorist attack is how to recover quickly. According to Robert Schainker, EPRI's Senior Technical Leader, "The physical hardening of some utility and power facility sites will be implemented with the help of the Department of Homeland Security, but utilities themselves need to prepare for responding and recovering effectively from the attacks. Utility-industry preparations need to include, at a minimum, vulnerability assessments; hardening of cyber communications and physical facilities; expansion of the store of high-voltage transformers and other long-lead-time equipment; and preparations for emergency transportation and installation of high-voltage transformers and other key equipment at utility sites."

FEDERAL AGENCIES

There are approximately 50 federal law-enforcement agencies in the United States. The majority of them are focused on urban America, with not nearly as much attention paid to our rural communities. Major agencies such as the FBI and the Secret Service have field offices close to rural America. However, these services have limited staff members to give the utility/energy industry the attention they need. Federal prosecutors are located in all states, but they experience heavy workloads and their availability is limited.

The Environmental Protection Agency (EPA) is one of many that offer important information for utility/energy infrastructure security.

Working together with the EPA, water utilities have ramped up security efforts at water-supply systems throughout the nation. Background checks on new employees have become common, as have intensive employee training, security audits, assessments, and emergency response and communications plans. A nationwide information-sharing system has been developed for water utilities. Utilities are identifying their most vulnerable traits and are working with local emergency first-responders to coordinate planning.

Taken together, this mobilization of effort and resources is virtually unprecedented. It has resulted in the development of:

- The EPA's "Baseline Threat Report" describing likely modes of terrorist attack and outlining the parameters for vulnerability assessments by community water systems. This is sensitive information provided only to water utilities.

- Risk-assessment tools for utilities to identify and evaluate their own security risks. Such analyses, called vulnerability assessments, are required by the Bioterrorism Act.

- Training programs on vulnerability assessments, used by several thousand water systems, to help utilities prepare accurate and detailed assessments.

- Security protocols to assure that vulnerability assessments are safeguarded after they are sent to EPA, as required by the Bioterrorism Act.

- Guidance and technical assistance for utilities to use in revising emergency-response plans as required by the Bioterrorism Act.

- Development of information on "best practices" and technical assistance on matters such as security hardware technologies
- An inventory of past security threats to community water systems and the lessons learned from them
- Analysis of the lessons learned by community water systems through the vulnerability-assessment process
- Guidelines that water utilities may use to guard against terrorists and security threats, correlated with the Department of Homeland Security's color-coded advisory system
- The Water Information Sharing and Analysis Center (WaterISAC), which provides a secure portal for the communication of sensitive security information among utilities, law-enforcement, and intelligence agencies

STATE AUTHORITIES

All states have state police, state crime bureaus, and prosecutors. However, staffs are limited and responding to rural areas takes time. Police departments and related state associations of law-enforcement personnel, however, offer a variety of support for business and industry no matter where they are located.

Although not formally part of the Criminal Justice System, other state agencies have resources and information that utility/energy facilities can use. Missouri's work with the National Guard offers just one of many examples. Recently, that state surveyed natural-gas operators throughout the state, providing them with vulnerability-assessment forms. More than half of the gas operating companies that responded stated that they might need assistance in planning by the National Guard.

Specifically, this study from Missouri states, "Of the approximately 90 operators contacted, vulnerability-assessment results were received from 13 companies including the major natural-gas operators in the state. Of the companies that responded, eight may have a facility that has a rating that might warrant additional consideration, evaluation, and review by the National Guard for inclusion in the State's safeguarding management plan."

Among the conclusions of this study is the following. "We recommend that the natural-gas operators be encouraged to review key facilities to determine if vehicle barriers might be helpful in making such facilities less vulnerable to terrorist attacks and to review their storage procedures for chemicals, such as odorants, to determine if additional security measures should be taken to ensure that such chemicals are not improperly used by terrorists."

In return for the help that your state's law-enforcement agencies offer all business and industry and for the assistance they willingly provide to your facility, consider giving financial support to law-enforcement agencies at state and local levels. You can support their associations by:

- Providing funds that help them secure information.
- Sponsoring additional patrol officers and agents.
- Helping them purchase needed equipment.

Police departments all over the nation are developing innovative ways to deal with terrorism. Here's a sample of progress being made in Maryland, Massachusetts, and New Jersey. A $7.5 million Home Security Initiative in Maryland is giving police cruisers statewide upgraded computers and two-way radios so that officers have important information about cars they are approaching before leaving the police car. In addition, the cruisers have in-car cameras that record all traffic stops. This new system provides access to the FBI's most-wanted list and to everyone on the state-police watch list.

In September 2001, state police in Maryland pulled over a suspect but, lacking appropriate information, let him go. Turns out he was among the 19 who perpetrated the 9/11 terrorist attacks. If they had this information-technology upgrade in their cars back then, they could have prevented him from joining the attackers.

New Jersey has had a Counter-Terrorism Bureau since 2003 to coordinate antiterrorist initiatives and to identify, deter, and detect terrorist-related activities in the state. This bureau works closely with the state's Counter-Terrorism Unit and the New Jersey Officer of Counter-Terrorism. As a group, they conduct investigations and develop intelligence information. Their purpose is to develop strategic intelligence assessments that identify domestic and international threats and refer them for further investigation.

New Jersey has two additional antiterrorism groups. The office of Counter-Terrorism is a civilian investigative unit that works cooperatively to identify and disrupt the ability of terrorist networks that operate within the state. And the Joint Terrorist Task Forces conduct investigations under the direction of the FBI. Members of these task forces have "top secret" national-security clearance and are sworn as deputies of the United States Marshals Service.

A different innovative program in New Jersey trains state police in the best ways to detect suicide bombers. State police, for example, watch transportation systems closely. Massachusetts offers the latest intelligence to law-enforcement officials across the state and around the nation, from the federal government down to the smallest police department. Termed the Fusion Center, it gathers and disseminates intelligence with the goal of becoming a nexus with local police and public-safety personnel. Its database includes terrorist information plus crime and intelligence analysis.

LOCAL POLICE

Most cities and towns within any state have police departments and city prosecutors. The remaining rural areas receive police services from sheriffs' departments. Most local police departments hold membership in the International Association of Chiefs of Police (IACP). The strategic plan of the IACP with local law enforcement guides policy and provides direction for the association. The Strategic Planning Committee, which is composed of members of the IACP Board of Officers, meets regularly to review the plan and reevaluate the goals and objectives in light of current and emerging issues in the law-enforcement environment.

Recently, IACP leadership revised the strategic plan, placing renewed emphasis on:

- Integrity and ethical behavior in professional policing
- Tolerance and appreciation for diversity within police agencies and with the public
- Partnerships and coalition building with the private sector and community groups
- Education about and acquisition of new technology
- Expanded research efforts, especially regarding youth and violent crime trends
- Community safety programs, especially traffic-safety efforts
- Enhancement of IACP's international outreach

Limited resources, rising crime trends, and geographic barriers can severely impair the ability of rural law-enforcement agencies to effectively combat crime in their communities. To ensure that they can meet the challenge of law enforcement in the 21st century, rural executives are increasingly seeking opportunities to provide more education and training for their staff and to maximize their effectiveness through new and improved technology.

The National Center for Rural Law Enforcement (NCRLE), a division of the Criminal Justice Institute, University of Arkansas System, is a university-based organization committed to helping rural law-enforcement agencies meet this challenge. One simple statement serving the needs of rural law enforcement comprehensively describes the important function that the NCRLE provides to help strengthen the law-enforcement profession in America's small towns and rural communities. By utilizing innovative strategies and a staff of experienced professionals, the center offers educational and training opportunities, research, technical assistance, and technical integration services customized to meet the needs of rural law-enforcement agencies across the country.

Since its inception in 1995, the NCRLE has developed a close working relationship with the Department of Justice. This relationship has enabled the center to work on a variety of law-enforcement initiatives, ranging from making law enforcement technology more accessible to rural agencies to battling problems in society, such as sexual assault and school violence.

NCRLE provides Internet access at no cost to rural law-enforcement agencies. Not only does this project provide Internet access and email; it also offers the following services to the law enforcement community:

- Free law-enforcement-agency website hosting
- Toll-free access to a help desk for technical assistance: 1-888-411-1713
- Hands-on assistance from the help desk in applying for the Bullet Proof Vest Partnership grant program, which is available online through the Department of Justice
- Online informational resources and publications; including model policies and procedures for law-enforcement agencies, detention facilities, and public-safety telecommunication centers
- Membership in CopShare, an email listserv that disseminates law-enforcement-specific information and successfully promotes the sharing of information between law-enforcement agencies nationwide

EMERGENCY PERSONNEL

Many states have emergency-services personnel (fire, ambulance, health-care). Most rural areas, however, have volunteer services. A number of major utilities have put in place their own fire departments and ambulance services due to budget constraints on volunteer departments. The National Emergency Management Association (NEMA) is a nonpartisan, nonprofit 501(c)(3) association dedicated to enhancing public safety by improving the nation's ability to prepare for, respond to, and recover from all emergencies, disasters, and threats to the nation's security.

NEMA was started in 1974 when state directors of emergency services began to exchange information on common emergency-management issues that threatened their constituencies. State directors of emergency management are the core membership of NEMA. Membership categories also exist for key state staff, homeland-security advisors, federal agencies, nonprofit organizations, private-sector companies, and concerned individuals. NEMA became an affiliate organization with the Council of State Governments (CSG) in 1990. CSG supports NEMA by providing an information and support network among state directors of emergency management, as well as interfacing with other national and regional organizations that provide emergency management information and state government policy related to emergency management.

When disaster strikes, coordination is needed among firefighters, paramedics, police, hospital personnel, government officials, and workers at emergency shelters. More and more communities are relying on the expertise of emergency-services managers to coordinate the myriad of emergency workers and volunteers. They help to ensure that government, volunteer, and medical personnel work together cooperatively and competently during a community emergency. Here is a list of different types of personnel that might be called upon during a security emergency, as well as a brief description of their roles:

- Emergency program managers and directors coordinate and advise city and state agencies and organizations. They develop special projects to prepare for emergencies and provide training in emergency management. They supervise staff and volunteers; direct the preparation for and response to natural and technological disasters; develop emergency plans, policies, and procedures; monitor emergency conditions; and advise public officials. They can work at either the city, county, state, or federal level and are often called by other titles, including emergency-services managers, emergency planners, and emergency-management specialists.

- Directors of homeland security specialize in developing counterterrorism programs.
- Emergency-operations-center chiefs manage emergency centers and command posts for private companies and government agencies.
- Emergency-preparedness instructors provide hands-on emergency training for private companies, emergency personnel, and volunteers.
- Risk-management experts develop programs to prevent industrial accidents at companies handling hazardous materials.
- Technical training supervisors develop, research, review, and maintain training programs in various aspects of emergency preparedness, such as antiterrorism or hazardous-materials emergency response.
- Hospital coordinators make sure that hospitals work together during an emergency. They create hospital plans and coordinate emergency response. They also oversee the distribution of funds to hospitals and other healthcare providers.

EFFECTS OF THE DISASTER EXPERIENCE

Most of the information in this book is straightforward. During disaster experiences, however, a variety of psychological problems can occur. It's important to know about and be prepared to deal with them. These problems can include:

- Emotional reactions: temporary (for several days or a couple of weeks) feelings of shock, fear, grief, anger, resentment, guilt, shame, helplessness, hopelessness, or emotional numbness (difficulty feeling love and intimacy or difficulty taking interest and pleasure in day-to-day activities)
- Cognitive reactions: confusion, disorientation, indecisiveness, worry, shortened attention span, difficulty concentrating, memory loss, unwanted memories, self-blame
- Physical reactions: tension, fatigue, edginess, difficulty sleeping, bodily aches or pains, startling easily, racing heartbeat, nausea, change in appetite, change in sex drive
- Interpersonal reactions in relationships at school, work, in friendships, in marriage, or as a parent: distrust; irritability, conflict,

withdrawal, isolation, feeling rejected or abandoned, being distant, judgmental, or over-controlling
- Stress reactions can range from mild, fleeting responses to disaster experiences to lasting posttraumatic stress disorder (PTSD), anxiety disorders, depression, intrusive re-experiencing of terrifying memories, nightmares, flashbacks, anxiety, and depression.

PTSD is a psychiatric disorder that can follow the experience or witnessing of life-threatening events. These traumatic events can include military combat, natural disasters, terrorist incidents, serious accidents, or violent personal assaults, such as rape. People who suffer from PTSD often relive the experience through nightmares and flashbacks, have difficulty sleeping, and feel detached or estranged. These symptoms can be severe enough and last long enough to significantly impair the person's daily life.

PTSD is marked by clear biological changes and psychological symptoms. And PTSD is complicated by the fact that it frequently occurs in conjunction with related disorders, including depression, substance abuse, problems of memory and cognition, and other physical and mental health issues. The disorder is also associated with impairment of the person's ability to function in social or family life.

After a disaster, here are some actions to take that can help those involved to cope better:

- Attend a debriefing. If one is not offered, work with local emergency personnel to organize one within two to five days after the disaster.
- Encourage everyone to talk about feelings as they arise.
- Be a good listener to your coworkers.
- Don't take anger personally. It's often an expression of frustration, guilt, or worry.
- Give your coworkers recognition and appreciation for a job well done.
- Eat well and try to get adequate sleep in the days following the event.
- Maintain as normal a routine as possible, but take several days to "decompress" gradually.

After the event, here's some important advice for everyone involved in security operations. When returning home:

- Catch up on your rest. This can take several days.

- Slow down. Resume your normal pace over time; don't rush things.
- Talk about the disaster or don't talk about it, whichever is right for you. Realize that people who haven't been through it might find the details frightening.
- Expect disappointment, frustration, and conflict. Sometimes coming home doesn't live up to idealized expectations.
- Don't be surprised if you experience mood swings. They will diminish with time.
- Don't just focus on yourself. Be sure to talk with any children in your life about what happened in their lives while you were gone.
- For more good information, visit
  http://www.ncptsd.org/topics/em_svcs_personnel.html.

EXPAND YOUR SECURITY PROGRAM

If your utility has facilities, substations, and infrastructure that extend through a region or state, your security team needs to develop and practice

Figure 6.2 Warning sign at a power substation

infrastructure-security programs that can monitor all facilities, even those that are the most remote. One of the great benefits of hardening targets is that the increased security sends a message to everyone who would do you harm. That message is: Try a softer target.

So if you have remote or dispersed sites that do not have enhanced security, be sure to monitor them carefully and beef up their security components as soon as possible. One of the ways to increase security while controlling costs is to work closely with other utility/energy groups in your region. Have your security team work with fellow utility-energy companies to develop a security collaborative for your region, perhaps the entire state, too.

With your security team in place, your security-assessment reports in hand, and your crisis-management plan not only written but tested, you have much to offer your colleagues. Find out what they can offer you.

Guidelines During and After an Incident

If, despite the increased security your facility has implemented, there is an incident, your security team and all your employees should be prepared to act. The security team will need to:

- Notify local, state, and federal authorities of the incident.
- Immediately implement the crisis-management program.
- As soon as prudent, notify local media so that they can stay in touch with your spokesperson, not people who lack authorization to speak.
- Throughout the incident, keep your lines of communication open with the criminal-justice system.
- Take every precaution to protect life and property.
- To prevent further accidents and injuries, do everything possible to secure the areas of the incident.

After an incident has occurred, your security team will want to further secure areas that have been affected by the incident. This action will both prevent additional injuries and the destruction of evidence related to the incident's origin. You, the security team, and any employees with relevant information will want to aid the criminal-justice system in the ongoing investigation. Be sure to have members of the security team document every aspect of the incident.

Chapter 7

SECURITY TECHNOLOGY

The security technology that I'll cover in this chapter includes current, little known, and new solutions that seek to protect utility energy facilities and their computer systems. The discussion will extend to current requirements and point to future technologies.

CURRENT UTILITY SECURITY TECHNOLOGIES

Let's begin with a major problem that is well known within the world of cyber security. Windows is the predominant operating system used on business desktops worldwide. It is also the dominant operating system used as a remote-terminal-unit (RTU) solution in many of the modernization efforts on supervisory-control and data-acquisition (SCADA) systems. However, according to Frank Earl, Senior Engineer, Coollogic, Inc., Dallas, Texas, "Contrary to Microsoft's protestations, Windows is by far the most *insecure* operating system in current use." So what do we do about this problem?

Windows Security

When most people think about computer security, they think of keeping intruders from hacking into a system or a network and, to offer one more example, keeping end users from seeing confidential information. Although computer security does cover these issues, it also involves much more. You need to protect your computers and the important information they contain by upholding a security-policy framework that supports technology. Your policies could include:

- Internet/Intranet security that explains appropriate use policies - These should also identify the types of prevention measures the organization is taking for spam blocking, adware and spyware, use of firewalls, and other security technology.
- Computer viruses and their prevention
- Network and file security, including protocol for administrator rights on systems and audit trails
- Network security
- User names and passwords
- Disabling all factory default backdoor passwords
- Workstation, portable computer, and personal-digital-assistant (PDA) security
- Remote access provisions, such as VPN and dialup
- Stringent physical security requirements, including lock-and-key control or card access for data-processing facilities and telecommunications/server rooms
- Resilience to non-login attacks
- Hardening your system so it's resilient to denial-of-service attacks
- Responding fast and cleanly to software and hardware updates

If you can't safely update or fix security issues, or the improvements take a long time in coming, your system is not adequately secure. According to Frank Earl, "Windows does not do a good enough job of keeping unauthorized users from accessing the system. It will keep attackers from logging into the main user interface, but it doesn't keep them from gaining access as an administrator, injecting controlling code that has the privileges reserved for the operating system itself, allowing someone to control the box by remote, or resisting non-login attacks."

To support his claims, Frank Earl cites viruses, worms, and significant spyware attacks that include "Slammer," "Nimda," and "Code Red" as evidence that Windows is not very resilient to non-login compromise attacks. In fact, al Qaeda operatives have been caught with information about the vulnerabilities of computer systems, SCADA systems, as well as structural diagrams of major dams, the power grid, and more. If they don't have the expertise themselves to perform attacks on these systems, they have enough money to buy help.

Yet there are some commentators who point to the overall complexity of the Internet and corporate computer networks as being actual deterrents to attacks. They say that cyber warfare is more the mutterings of network-security companies having something to sell than it is a real threat. But what if there were a month-long blackout over an area as large as the 2003 blackout? What if it covered the entire country in a breakdown of telecommunications or power that lasted several months? What if you had a mass failure of your RTUs at all your facilities, causing dangerous conditions for each location? In many cases this is entirely possible because of weaknesses and vulnerabilities of the SCADA systems in general combined with weaknesses within Windows.

Mitigating the Risks

What can we do to decrease the risks we've talked about? Here are some initial actions:

- Download the *latest* version of Windows on *all* machines in the enterprise. While Windows isn't the best choice, if you have primarily XP and 2003 servers in place, short term it doesn't make much sense to replace them with something completely different right away. If you have a Windows 2000 or Windows 98 box in the mix, you're at very high risk for an attack. Even then, you really should make a plan to get away from Microsoft's offerings until they show signs of offering better security in at least the medium-term future.
- Use a topnotch antivirus tool. AntiVir (http://www.antivir.de) and MacAfee (http://www.macafee.com) are two good choices. They work well for users and catch problems more often than most other products.
- Install an effective spyware removal tool. AdAware (http://www.lava-soft.com) and SpyBot Search and Destroy are two of the top tools by user ratings and shootout comparisons.

- Perform regular signature updates for these programs
- Use monitoring software, such as SNORT, which is free, and SourceFire's 3D monitoring system, which is SNORT combined with several proprietary add-ons that dramatically improves its detection ability
- Install all current security fixes, including Windows XP's Service Pack 2
- If possible, upgrade *all* machines to Windows 2000/XP and Windows Server 2003.
- SNORT, a wireline intrusion detection system for networks, is one of the best tools for this job. (http://www.snort.org) and Sourcefire (http://www.sourcefire.com). Commercially supported SNORT (the original authors of SNORT founded this company) with enhanced monitoring and response (as in modifying firewall settings to block off attackers, etc...)

Please note that the tracking of each new virus and worm can lag as much as four weeks behind the initial introduction into the world of each new type of malware. In other words, a worm or virus can still blast through a corporate or SCADA network without initial notice. All of which brings us back to the vital subject of upgrading the security of our SCADA systems.

Improving SCADA System Security

There are many threats to SCADA systems. Some of them stem from being run on a Windows-based platform instead of something more robust. Some problems result from SCADA's relatively simple implementation, which was designed with too little concern for overall system security. Other problems related to SCADA include bugs in the software that in turn cause a loss of control of the system being managed and monitored by SCADA.

One of the traditional ways to protect SCADA systems and corporate networks is by providing rings of defense. This strategy involves the analysis of multiple layers of both the corporate network and the SCADA architectures, including firewalls, proxy servers, operating systems, application-system layers, communications, and policies and procedures. It's best if your strategy for SCADA security complements the system in place to keep the corporate security network secure. Additional approaches for securing SCADA systems are:

- Put cryptographic modules on the endpoints for communication
- Reactive responses to overall Windows issues
- Careful and constant monitoring
- Install secure gateways acting as intermediaries to a legacy SCADA system

"We offer monitoring solutions and strongly recommend their use, but we also strongly advocate proactive measures, including minor to major redesigns of the SCADA plant that are intended to prevent most exploits by removing the attack vectors completely," says Coollogic's Frank Earl. There are two current frontrunners in the arena of configuration management (CM) to be placed in between the remote terminal units (RTUs) and the network operations center (NOC):

- The American Gas Association's recommendation number 12, referred to as AGA12 and having a default implementation provided as open-source software called "scadasafe," is a Java-based software solution.
- Honeywell's "beepbeep" algorithm and Electric Power Research Institute's (EPRI) suggested solution, which is similar to AGA12.

As mentioned above, the overall security of the SCADA system is relative to all the components of the SCADA system. Most security measures that are being used to date tend to focus on the communications pathway. This assumes that, if you deny an aggressor the ability to communicate with the systems by remote access, you are providing the most overall security with the least amount of effort and expense. The problem is that it's not just the RTUs you need to worry about. It's the components of the system as a whole.

Unfortunately, there are other ways to subvert a system with low effort or to deny positive control of the system and get the same result as if you'd simply hacked into a substation's control systems. All communications security solutions rely on having a relatively high level of trust among all the SCADA components. But what if you can't trust some or all of the components within the system? To heighten the security of your SCADA system:

- Rework as much of the SCADA system's infrastructure as possible
- SCADA RTUs should *not* be directly reachable from the corporate network.

- SCADA systems should have active password protection *above and beyond* what the corporate network's access provides.
- Dialup modems should be discarded where possible. Where not possible, rework the dialup models so security doesn't rely on obscurity. Consider using a dial back modem.

Here's a special note about SCADA systems with proprietary protocols. Many SCADA systems utilize some sort of proprietary communication protocol between the NOC system software and the RTU system software. Please understand that these protocols do *not* secure the SCADA from attack. Why not? Because these protocols can be reverse-engineered, and some of the formerly closed detail protocols have been opened in an attempt to establish them as standards. As a result, the opening enables anything that's not properly secured to be attacked by other means.

Anti-Spyware and Anti-Virus Solutions

There are no absolutely sure ways to prevent spyware from being installed on a Windows system. But here are some of the best ways to mitigate that risk:

- Install Mozilla (http://www.mozilla.org), Mozilla Firefox (next generation of Mozilla/Netscape that is currently in development), or Opera for browsing and Pegasus Mail, Mozilla Thunderbird, or some sort of web mail application for regular e-mail.
- You can install Opera (http://www.opera.com), a web browser that lighter than Internet Explorer in code size but not quite as standards-compliant as Firefox. There is a charge for Opera, but Firefox is free.
- Never allow any software to be installed by end users.
- The information-technology (IT staff) should carefully inspect any installations for spyware being injected by a chosen application.
- AdAware is an excellent after-the-fact tool to deal with spyware. It catches and removes most of the known instances with relatively few exceptions. SpywareRemoverKiller is another good tool.

Encryption

SSL-based solutions are relatively good, but they rely on the current public key infrastructure (PKI) that is provided by such companies as Thawte and Verisign. These solutions are well known and relatively secure, but they rely on a third party for authenticating the cryptographic certificates. This is a relatively good means for securing website activity, yet it leaves some vulnerability in a SCADA system's outlying nodes. Other tools use their own authentication, and therefore they are only as secure as the authentication scheme and the choice and integration of the cryptographic algorithms within the main protocol. Good strong cryptographic algorithms help considerably, but mistakes can be made that weaken the security.

Encryption provides an increased level of security, but the endpoints are a problem. Cryptographic modules prevent a person from snooping on the conversation in the middle on the wire. It does nothing to prevent someone from hijacking the controls at the RTU end by an initial physical attack to provide an end run around the CM's. It does nothing to prevent someone from attacking the NOC end of the equation either—the end-user workstations are vulnerable to attack. Although CM's are a good quick-fix start to securing systems, they are not the be-all, end-all solution that others have made them out to be.

Firewalls

Firewalls are part of the comprehensive solution to the many vulnerabilities of Windows and SCADA systems. Here are some of the components of firewalls:

- Dialup modems are placed at the RTU end of the equation. The RTU side should be calling in, not the other way around. This way you can't remotely attack it through that vector.
- Dial back modems, placed where you connect, provide a user ID/password and call back to the known phone number for that user ID/password. These modems offer small protection against remote attacks, but they are likely to be ineffective against a sophisticated attacker.
- Modems in general can be fooled. A sophisticated person can wedge in on the copper pair and force off the NOC side, thereby taking control. Modems are easier to attack and subvert in ways that are largely undetected. They tend to be less secure than mobile-phone modem setups and TCP/IP-based communications.

  The RF modem or wireless link setup is a type of dialup modem that is vulnerable to attack. If you use either or both of them, make alterations to the system to prevent subversion by this vector. Secure the remote end by not allowing any control or data collection to occur against the RTU directly.

- In cases where corrective actions are not possible, the dialup system can be re-worked to establish a session with the NOC's modem bank through a modem installed on the same communications gateway. This strategy establishes a TCP/IP-like connection back to the NOC. Note that in remote locations lack of bandwidth might make this alternative impractical.
- Contractor lines are a backdoor way for a contractor to work on systems or for the supplier to work on the RTU by remote. For security reasons, contractor lines should always be ripped out of the system when the contractor's work is done.
- Virtual private network (VPN) laptops: SCADA systems should *not* expose control aspects to the corporate network except through the SCADA user interfaces. The SCADA system needs its own user accounts and logins, above and beyond those belonging to the corporate network. These should only be stored on the SCADA system itself.

The most important action is to change the design of the SCADA system so that someone can never directly talk to the equipment at the site. Instead, access needs to be managed at all times by a cryptographically secured intermediary.

One of the primary ways to protect your facilities and infrastructures is through the gathering of video intelligence. To accomplish that, you want to use systems that are covert, portable, built to withstand a variety of weather conditions, and able to incorporate both manned and unmanned surveillance. The following products can be used to enhance existing static surveillance systems, and they can also provide interim surveillance when there is no system present (I'm grateful to Brian Jacobs of Surtec (www.surtec.com) for his advice in compiling this information):

- A fiber optic intrusion system for perimeter fences and walls
- The LM-BOS combined with the covert cameras. The active motion-sensing capabilities are combined with the overall hostile environment design of the system up front.
- In addition, it is possible to wire in some of the covert 110 cameras into the lighting of the facility. The DRS-5 would be good for providing an interface and log of interrogations using covert cameras in clocks and other non-obvious camera placements.

NERC Meets FERC Security Mandates

The North American Electric Reliability Council (NERC) has developed cybersecurity standards for voluntary compliance by the electric industry. Termed NERC 1200, utilities are still uneven in following these standards. A new standard, NERC 1300, will soon supersede NERC 1200, as the newer version extends coverage beyond distribution and transmission companies to also include power generators. NERC 1300 carries with it enforcement provisions and penalties.

The Federal Energy Regulatory Commission (FERC) wants to be sure that the voluntary standards are complied with. If not, FERC might react by imposing government regulations. If there's a blackout that is traced to noncompliance with NERC cyber-security standards, for example, FERC would likely contact the offending company.

FUTURE TECHNOLOGY REQUIREMENTS

The electric grid involves more than 3,000 companies, most of which are investor-owned utilities. Parts of the grid, however, are owned by federal agencies, rural electric cooperatives, and municipalities. Grid reliability is administered by about 150 control area operators throughout North America, and these operators are coordinated by some 20 regional reliability councils. The regional councils operate under NERC's operating and planning standards function only as guidelines, however, because they are all voluntary. There are a variety of interpretations and implementations of these voluntary guidelines, and these variations will continue until NERC standards become federal law.

Ever-Growing Demands

Meanwhile, demands for electricity are constantly growing and changing, causing our aging infrastructures to strain as they seek to meet the increasing demands placed upon them. Keep in mind that our power-delivery system is more or less based on technology developed more than fifty years ago. It was initially installed that long ago, too. The many challenges facing our nation's electric grid include:

- The rapid increase in wholesale transactions between, for example, independent power producers and distribution utilities
- Ever-increasing grid congestion

Figure 7.1 Steel poles support increasing need for electrical power

- Low levels of investment in the infrastructure
- Continual growth in technology that enables more alternatives for consumers
- The greater need for improved grid security for physical vulnerabilities, as well as disruptions to computer networks and communications systems

- The vastly increased electricity demands of our digital society
- Inconsistent and conflicting regulations
- Market reforms that have produced rules that vary from state to state and in some cases from utility to utility within certain states
- Federal open-access orders that demand open competition yet fail to give clear direction about the ways open competition must be implemented.
- Changing environmental policies

The Need for More Electricity

According to the Electric Power Research Institute (EPRI), there is a great disparity between the expected growth in demand for power and the expansion of the delivery system to meet that demand. In "Electricity Technology Roadmap: 2003 Summary and Synthesis," EPRI states that it expects demand during 1999 through 2009 to grow by 20 percent but system capacity to increase by only 3.5 percent. That's a consequence of low levels of infrastructure investment.

The constantly growing demands of digital technology extend way beyond desktop and laptop computers. Computers are embedded in home appliances, industrial sensors, and so many other devices that there are more than 12 billion in our nation alone. In fact, the ratio of microprocessors in standalone applications compared with those inside computers is 30 to 1.

All of these devices are highly sensitive to variations in power quality and disruptions of even an instant can impair performance. Currently, about 10 percent of the total electric load in the United States serves digital technology. The "Electricity Technology Roadmap: 2003 Summary and Synthesis" projects that the digital-quality power load will reach 30 percent by 2020. It could even go up to 50 percent.

Yet our electricity infrastructure was designed in the 1950s to serve analog—or continuously varying—electric loads. There is no way it can deliver the needed level of digital-quality power our growing dependence on microprocessors requires. All of which makes our nation's electric infrastructures even more vulnerable to terrorist attack. The great strain on our grid means that even minor disruptions can lead to cascading outages.

That's what happened during the middle of August in 2003, resulting in the 2003 blackout. This repeated, uncontrolled cycle of overload leads to equipment failure that in that case affected 50 million Americans and Canadians.

Near-Term Needs

EPRI's Infrastructure Security Initiative (ISI), which was launched in the wake of 9/11, is developing pilot-scale prototypes of new hardware and security assessment tools. The ISI is working with several federal agencies, including the U.S. Department of Homeland Security and the Department of Energy, to coordinate EPRI's research and development with government-sponsored planning and technology development. Rapidly deployable, high-voltage recovery transformers are needed and are in the design stage. Power-equipment manufacturers ABB and a joint venture of Mitsubishi and Toshiba are readying different prototype designs. These transformers have a three-year design life and a range of power ratings. This advanced technology could be pre-positioned for quick delivery and installation in response to terrorist attacks on multiple electric substations.

Figure 7.2 Wooden pole with transformers

Better Wireless Security Is Required

We currently rely on Federal Communications Commission (FCC) regulatory rules to protect non-wireless local-area-network (WLAN) wireless links. Some systems use rfmodem links back to other areas where they can be linked into leased lines, Public Switched Telephone Network (PSTN), Internet, and more. Most of the systems currently in use rely on the need to license such equipment with the FCC and the relative difficulty of the general population to obtain the equipment for protection. It is not good practice to rely solely on security schemes intended for regular home or business use through WLANs that attempt to secure the links being run on off-the-shelf WLAN hardware. Most of the security measures that have been developed for WLAN use have not truly secured even consumer WLAN networks, let alone SCADA systems. Currently there are at least three different security protocols that have been proposed and actively used for WLANs:

- Wire equivalency protocol (WEP)
- Lightweight extensible authentication protocol (LEAP)
- WiFi protection architecture (WPA)

WEP and LEAP don't offer enough security, and WPA has not been in use for a substantial amount of time, so its level of security remains to be seen.

Long-Term Needs

To support the digital society of the 21st century, we must transform our nation's electric grid into a smart power system. We must develop an integrated, self-healing electric grid with excellent resilience that responds instantly to the billions of decisions made by microprocessors. To build a more resilient, less vulnerable electric grid, here are the types of technology we need:

- Automation—In order for us to have reliable, high-quality electric service, we need automation throughout our grid. As a result in an emergency, distribution feeders and/or locally distributed energy resources will automatically be isolated. The automated grid would provide a self-healing, self-optimizing system that responds to disturbances, even anticipates them and reacts in ways that keep the grid running.

- Computer modeling–Retail power markets keep changing, so we need computer modeling to test new market models. In this way we can determine the right incentives to offer buyers and sellers that will expand the electric grid and protect it from increased risk.
- Distributed resources–The best way to increase the resilience and reliability of the electric grid, while maintaining low-cost and high-quality power, is to build small generation and storage devices. These resources will be distributed throughout the nation and seamlessly integrated with the grid.
- Energy and communications architecture–Before we can build a truly smart electric grid, we need communications architecture that is standardized and laid over today's system. Publicly available standards should be the basis for any integrated energy and communications-system architecture.
- Solid-state controllers–These devices have the speed and accuracy of microprocessors at a level of power that can be 500 million times greater.
- Superconducting cables–We need a variety of new solutions to carry far more power through current transmission corridors. Some solutions will involve new materials using, for example, carbon-fiber cores. Other solutions include reconfiguring existing cables and developing electric conductors that weigh less and function better at the high temperatures that accompany high flow rates. High-temperature superconducting cables that are suitable for underground systems are likely to carry more than three times the current of current conductors.

As you have probably guessed, these necessary advances will cost billions of dollars. But investments in the increased security of our nation will likely not be made in whole, and they'll only be made in part in a long, drawn-out manner. Unless, of course, we get another major terrorist attack. Then we will react fast. But for thousands or tens of thousands of Americans, it will be too late. We must move from a reactive to a proactive strategy. That is the only way to make our nation safer.

Chapter 8

INDUSTRY-SPECIFIC
SECURITY ISSUES

THE BIGGEST THREAT

We're all interested in certain fundamental threats. The very biggest is the possibility of a dirty nuclear bomb being exploded within our country, most likely in a major city. A dirty nuclear bomb is a conventional explosive laced with radioactive material. Estimates of the economic cost of a crude nuclear device detonated within a major population area run into the trillions of dollars. Beyond that, there could be hundreds of thousands of deaths, a zone of destruction measuring one or more miles wide, and radiation extending from the center of the blast for several miles.

More than 2,000 metric tons of nuclear materials exist around the world, yet only several kilograms are needed to construct a bomb. Financial and diplomatic pressures can theoretically be applied to control access to nuclear material and therefore prevent the building of nuclear bombs. Outside the United States, most of the world's nuclear material is located in the former Soviet Union. America and other members of the G-8 countries have pledged $20 billion to reduce the threat of nuclear terrorism, although not all of that money has been allocated.

In addition to the countries in the former Soviet Union, some 40 other nations have nuclear weapons or materials that could potentially be sold to terrorists or stolen by them. And many of these countries have insufficient security surrounding these deadly materials. In addition, some of these nations have unstable governments. North Korea is one of them, and it certainly has the potential and mindset to equip terrorists with nuclear materials.

Terrorists groups do not have the ability to produce nuclear material, so at least the nuclear inventory is finite. We need the will and resources to control it, however, all over the world.

THE ELECTRIC UTILITY INDUSTRY

Unlike natural gas and coal, electricity is generated as it is used. There is very little ability to store it. So the electric power system must constantly be adjusted to assure that the generation of power matches its consumption. About 150 control-area operators serve this function within the continental US, and they use computerized control centers to dispatch generators as needed. To help the control-area operators, electrical generators are divided into three main categories:

- Baseload power plants. They run all the time to meet minimum power requirements.
- Nuclear power plants, for example, are almost always operated as baseload plants, because these plants are most stable at full power.
- Peaking power plants. They only run to meet the power needs at peak loads. These are often the most expensive plants on the system to operate. Often, they are small, older coal-or oil-fired plants, although gas turbines are sometimes used as peaking plants.
- Intermediate power plants. These facilities fall between baseload and peaking power plants and are used to meet intermediate power loads. Intermediate plants adjust to changing power. Gas turbines can be used as intermediate plants.

Although control-area operators run the grid within their control areas, the responsibility for electric grids rests with electric utilities.

Figure 8.1 LaCygne

They operate the electrical grid within their service area by coordinating the efforts of the control-area operators in their service area. Electric utilities are also responsible for investment in new lines, maintenance, and control of access to the grid.

The competition in wholesale and retail power markets has caused some people to view utility control of the grid as a conflict of interest. Some states have moved to pass the control of the grids to independent system operators (ISOs). They function in California, Texas, and New England. Ownership of the transmission and distribution systems is often retained by the utilities, but it can be passed off to independent transmission companies, often called TransCos. In the latter case, the utility effectively becomes a distribution company, called a DisCo.

How the Electric Grid Works

Traditional models of electric-power generation and delivery are based on large, centrally located power plants. "Central" means that, ideally, power plants are located on a hub surrounded by many electrical load centers. A power plant might be located close enough to serve the electrical loads in the city and its suburbs or at the midpoint of a triangle formed by three cities.

Wherever power plants are located, their power must be brought by the plants to the users. That's the purpose of the electric grid. The system actually consists of two separate infrastructures:

- The high-voltage transmission systems, which carry electricity from the power plants and transmit it hundreds of miles away
- The lower-voltage distribution systems, which distribute electricity to individual customers

The transmission system is the central trunk of the electric grid. Thousands of distribution systems branch off from this trunk into tens of thousands of feeder lines reaching into homes, buildings, and industries. The "grid" refers to the transmission system, which truly is a grid. Transmission lines run not only from plants to load centers but also from transmission lines to other transmission lines. This strategy produces a redundant system that helps to assure the smooth flow of power. When a transmission line is taken out of service in one part of the power grid, electricity can be rerouted to continue delivering power to the customers.

The power grid offers an extremely inviting target. The U.S. grid is more of a prime target than power grids in other countries largely because in the United States, electricity drives computer networks that are the backbone of everything from schools to traffic-control devices to government agencies. Without electricity, the country would shut down.

The power blackout that struck the northern United States and Canada in August 2003, shocked industry executives into acknowledging the need to upgrade outdated circuitry and power-generating equipment. Some analysts now are warning, however, that the outmoded intrusion-prone computers used to control the grid pose a greater risk than the aging grid infrastructure. The patchwork system of computers loosely linking the electricity producers in the United States and Canada is an easy target for an Internet attack that could shut down the power grid.

Faulty Controls

For starters, the computer systems that run the maze of electrical-power grids are largely vulnerable because of poor monitoring. The oversight of

Figure 8.2 Substation

this equipment, which controls the connections to the power grid, must be constantly monitored. Believe it or not, most of these computer systems have not changed since the 1980s. There are numerous types of systems, and each one has its own set of rules. Inadequate network security measures are equally responsible for the high risk the electrical grid faces. Utility-company equipment is a mixture of networked Windows and Unix machines that are connected to the same Internet Protocol (IP) networks found throughout the Internet.

The nation's electric grid will not be secure unless and until we disconnect these utility-company computers from the Internet. This lack of control within the industry and ease of access from outside have led to a major concern about Internet-based attacks on the grid. One utility on its own, no matter how big or small, can make a significant improvement in the vulnerability of the nation's electric grid. Terrorists aren't the only threat to the electrical grid. Disgruntled workers and hackers also pose threats. Even innocent actions such as allowing a company employee to access the corporate network from home can open the door to security breaches.

THE OIL AND GAS INDUSTRY

The United States has 400 offshore platforms, 140 refineries, 170,000 gas stations, and 150,000 miles of liquid pipelines. We have 103 nuclear power plants, which produce 20 percent of U.S. electricity. Terrorist organizations with a global reach, very much including al Qaeda, have identified the world's energy system as a major vulnerability and a certain way to deliver a blow to America's oil-dependent economy. Terrorists can hurt us without coming to the United States, for example, by disrupting world oil installations such as pipelines, refineries, processing facilities, tankers, and oil-company employees.

The best way to protect the U.S. energy system from the threat of terrorism is to embark on a massive effort to reduce our dependence on oil in the transportation sector. Hydrogen fuel-cell vehicles are far off in the future, but hybrids and plug-in hybrids are already available, and if combined with capabilities to run on alcohol fuel, these alternative vehicles could significantly reduce our demand for oil. Alcohol fuels can be derived from biomass and from coal, both resources the U.S. and rising oil-consuming nations, such as China and India, have in abundance.

While we slowly increase our consumption of alternative fuels, we must rapidly expand our capabilities for accurate intelligence collection and thorough analysis. We don't have adequate intelligence because we don't have enough dedicated analysts and policymakers focused on energy security. We need a global effort that includes intelligence gathering, threat audits, and aggressive pursuit of terrorists.

For the oil and gas industry, supervisory-control and data-acquisition (SCADA) and industrial-control systems have tended to be proprietary, non-PC-based systems. They have escaped most cyber attacks because they use proprietary communication methods, for instance, to exchange information between the field and the operations center. These systems are moving to standardized, open technologies. While these changes have helped improve oil and gas operations, their openness has also made them more prone to security risks.

This and many other industries use Internet-protocol (IP) networks, which are open technologies. Because they can be implemented easily and interconnect with corporate networks, they allow business users to gain access to field data much faster. Unfortunately, their connection with corporate networks and the Internet also exposes them to many more risks of attack and sabotage through viruses, worms, and hacking.

DOE Tackles Security for Oil Industry SCADA Networks

The U.S. Dept. of Energy (DOE) has published 21 steps to help organizations improve the security of supervisory-control and data-acquisition (SCADA) networks. The list of action items includes:

- Identifying all connections to SCADA networks
- Disconnecting unnecessary connections to the SCADA network
- Evaluating and strengthening the security of any remaining connections to the SCADA network
- Hardening SCADA networks by removing or disabling unnecessary services
- Avoiding reliance on proprietary protocols to protect the system
- Implementing security features provided by device and system vendors
- Establishing strong controls over any medium that is used as a backdoor into the SCADA network
- Implementing internal and external intrusion-detection systems and establishing 24-hour-a-day incident monitoring
- Performing technical audits of SCADA devices and networks and any other connected networks to identify security concerns
- Conducting physical security surveys and assessing all remote sites connected to the SCADA network to evaluate their security
- Establishing SCADA "Red Teams" to identify and evaluate possible attack scenarios
- Clearly defining cyber-security roles, responsibilities, and authorities for managers, system administrators, and users
- Documenting network architecture and identifying systems that serve critical functions or contain sensitive information that require additional levels of protection
- Establishing a rigorous, ongoing risk-management process
- Establishing a network protection strategy based on the principle of defense-in-depth
- Identifying cyber-security requirements
- Establishing effective configuration-management processes
- Conducting routine self-assessments
- Establishing system backups and disaster-recovery plans

- Establishing expectations for cyber-security performance and holding individuals accountable for their performance
- Establishing policies and conducting training to minimize the likelihood that organizational personnel will inadvertently disclose sensitive information about SCADA system design, operations, or security controls

THE NUCLEAR INDUSTRY

Although North Korea and other nuclear-capable countries represent a possible threat from abroad, security measures taken at commercial nuclear facilities in our own backyard are helping to ensure America's safety. The Nuclear Regulatory Commission (NRC) and the nuclear industry have worked together to institute extensive new security measures. These measures protect U.S. nuclear plants from sabotage. The NRC requires nuclear power plants to take adequate measures to protect the public from the possibility of exposure to radioactive release from such acts by focusing on:

- The physical construction of the containment building for reactors
- Security personnel, procedures, and surveillance equipment
- Security-clearance background checks and daily monitoring for plant employees

What if terrorists attacked U.S. nuclear facilities? Here are some scenarios:

- A massive release of radiation after a nuclear plant is hit with a bomb delivered by truck or boat
- A 9/11-type attack using a plane as a guided missile to crash into a nuclear facility
- Sabotage at a nuclear facility by an insider or by intruders
- A ground assault on a nuclear plant by a commando team attempting to blow up the plant
- Suicide terrorists attempt to break into a nuclear plant and quickly build and then detonate a dirty nuclear bomb. Attackers could also use conventional explosives to blow up some nuclear waste or nuclear fuel, thereby spewing radioactive materials into nearby areas.

- Terrorists might target the pools in which nuclear reactors' highly radioactive waste, also called spent fuel, is kept. This waste, which is kept cool by water, could ignite if exposed to the air. The explosion could cause a catastrophic fire.

Note that all nuclear-power plants in this country are guarded by private security forces hired by the plants and supervised by the NRC. The result of the above list of terrorist attacks could not lead to a nuclear explosion. But the attackers could cause a meltdown, a major fire, or a major conventional explosion. All of these acts could spew radiation into nearby cities and towns.

THE TELECOMMUNICATIONS INDUSTRY

The worldwide growth of telecommunications networks, as well as their integration with the Internet and other open-network technologies, has led to a significant increase in universal connectivity. One consequence is that security in telecommunications networks has become a major concern at both the national and international levels. This concern results from the increase in potential damage that can be done to these networks from both traditional security threats that have not been adequately addressed and new security threats inadvertently created from the greater connectivity and new services.

Solutions have not been developed for such well-known security problems as source and destination authentication and communications integrity and privacy. The weaknesses in both wired and wireless network protocols have made popular denial-of-service attacks against the availability of network services tempting targets for many network attackers.

Here are some of the solutions developed to provide security for Internet-based networking:

- Firewall technology offers an effective starting point for access control in any distributed network.
- Virtual private networking (VPN) offers a secure and cost-effective alternative to leased lines. VPNs have been leveraged to fill the security gap in 802.11 wireless local-area networks (LANs). Designing, implementing, and integrating VPN technologies to existing infrastructures still require development.

- Research in security services for on-demand tunneling, ease of integration, interoperability, and security-policy enforcement across multiple domains continues to promise results.
- Other protocols for session-layer security, such as the SSL and its descendant TLS, might provide security services, such as confidentiality and data integrity for web-based client/server communications. These protocols, in combination with certificate technology for origin and destination authentication, are offering more benefits.
- Digital signatures and certificates can be leveraged to provide security services for non-repudiation in large-scale network applications. The deployment of lower-level security service, through IPSec and IPv6 protocols, will reduce many known inherent security weaknesses. It is not easy, however, to deploy them broadly.

The propagation of viruses and distributed denial-of-service attacks on a worldwide scale are well-known and significant threats. Wireless networking complicates routing and makes security difficult to achieve, as does ad-hoc networking, which provides network services without static infrastructure or centralized administration. In an ad-hoc network, there are no gateways, routers, or switches to forward packets, and routing is on a multi-hop peer-to-peer basis. The topology of these network is dynamic with nodes that are usually mobile and semi-controllable or randomly moving. Examples include military networks and dynamic-conferencing networks.

The lack of static infrastructure causes several security issues in the mobile-ad-hoc-network (MANET) environment, including node authentication and secure routing. Because the nodes in many MANET applications are located in a hostile physical environment, identifying compromised nodes is a task that requires new security technologies.

Secure routing in MANET applications is subject to several new security threats, such as black-hole attacks or routing-table overflow. In a black-hole attack, malicious or faulty nodes may pretend to provide the shortest paths for routing packets of other nodes. In a routing-table-overflow attack, the adversary creates routes to nonexistent nodes to cause denial of service to actual route requests due to overflowing.

The added complexity security brings to the network infrastructure creates an additional need for management and monitoring of the security components. Failures in and circumvention of the network infrastructure require an instant response mechanism against these attacks. Intrusion detection systems (IDS) provide mechanisms and protocols to detect incidents, collate data, and report security issues to network managers.

We are not adequately securing our telecommunications technology. We need to upgrade our existing systems with enhanced security services and products. Here are issues we as a nation must address in order for the telecommunications industry to move forward with increased security:

- Law-enforcement and public-safety officials need to get together with telecommunications and information companies to determine the best ways to detect telecommunications and other communications among terrorist organizations and other criminal groups.

- The Department of Defense (DoD) and the Cellular Telecommunications and Internet Association (CTIA) must settle their disputes over 3G and the rollout of spectrum-efficient broadband wireless and location services.

- Issues concerning spectrum caps need to be resolved so that the development of nationwide, spectrum-efficient infrastructures can proceed.

- Public switched telecommunications network (PSTN) needs to be made more robust so it is not as vulnerable to attack.

- The FBI, the U.S. Secret Service, and other law-enforcement agencies have criticized the telecommunications industry and the FCC for ignoring their attempts to deploy eavesdropping equipment into telecommunications infrastructures. We need to resolve important privacy issues as they conflict with vital security problems.

THE WATER INDUSTRY

Our water infrastructure includes surface and groundwater sources of untreated water for municipal, industrial, agricultural, and household needs; dams, reservoirs, aqueducts, and pipes that contain and transport raw water; treatment facilities that remove raw-water contaminants; finished water reservoirs; systems that distribute water to users; and wastewater collection and treatment facilities. Across the nation, these systems comprise more than 75,000 dams and reservoirs; thousands of miles of pipes, aqueducts, water-distribution, and sewer lines; almost 170,000 public drinking-water facilities (some serving as few as 25 customers); and more than 15,000 publicly owned wastewater-treatment facilities. Ownership and management are both public and private. The federal government is responsible for hundreds of dams and diversion structures. The vast majority of the nation's water infrastructure, however, is either privately owned or owned by local, regional, or state governments.

About 15 percent of the nation's large drinking water and wastewater utilities, located primarily in urban areas, provide water services to more than 75 percent of the U.S. population. These systems represent the greatest targets of opportunity for terrorist attacks. Our nation's reservoirs have long been recognized as being potentially vulnerable to terrorist attacks of various types, including physical disruption, bioterrorism/chemical contamination, and cyber attack. Damage or destruction by terrorist attack could disrupt the delivery of vital human services, threatening public health and the environment and possibly causing loss of life.

According to Tom Austin of IBG Security Consulting in New Hampshire, "There are compliance standards in the financial-services and healthcare industries that require certification. We need them in the water industry, too, but we don't have them. Security is not at the forefront of management's or Congress' mind, unless and until there's a big problem. When security contracts are let, they're made with engineering firms that, typically, know little about security."

To date, the Environmental Protection Agency (EPA)'s Office of Water has issued measures for drinking water and wastewater systems to guard against terrorist and security (DHS) threats. These measures are consistent with the Department of Homeland Security's color-coded advisory system, which begins with green, denoting low risk of terrorist/security threats, and increases in seriousness through blue, yellow, orange, and red. Water utilities are included in the 13 critical sectors identified by the DHS as potential targets of attack.

The EPA has taken a two-pronged approach in disseminating these precautionary measures:

- First, through the Association of State Drinking Water Administrators, each state's drinking-water program managers and staff receives guidelines and determines how best to coordinate these measures with other instructions developed by the state.
- Second, these guidelines have been posted on the Internet at a secure, password-protected site, available to drinking-water utilities only. The guidelines provide information to water utilities on a variety of critical water-infrastructure-protection activities.

Security Threats to Water and Wastewater Infrastructure

From time to time, U.S. officials increase the terror-alert status for the country from yellow (elevated) to orange (high), and we have a renewed focus on the vulnerability of water and wastewater plants. A report by the Environmental Law Institute (ELI), for example, highlighted the combined threats to U.S. drinking-water supply from terrorism, aging infrastructure, and other emerging biological and chemical contaminants and mutating pathogens. Environmental Defense (ED), to cite another example, has warned that chlorine gas used at wastewater-treatment facilities poses an unnecessary risk to communities from accidental release or even terrorist attack. Based in New York, this organization advocates replacing chlorine gas with less toxic liquid sodium hypochlorite or with ultraviolet (UV) treatment.

The physical threats to drinking-water and wastewater infrastructure are focused on treatment plants, pumping stations, pipelines, and disinfectant storage facilities. There is also a growing concern about the intentional dissemination of biological, chemical, or radiological agents through drinking-water distribution systems. Priority pathogens have been organized into Categories A through C by the Centers for Disease Control and Prevention and the National Institute of Allergies and Infectious Diseases. Examples of potential chemical contaminants include:

- VX
- O-Ethyl-S-(N 3 N-dimethylaminoethyl)
- Methyl thiophosphonate
- Sarin
- Colchicine
- Cyanide, amiton
- Fluroroethanol
- Sodium fluoroacetate
- Selenite
- Arsenite
- Arsenate

Conventional surface-water drinking-water treatment plants consist of multi-barrier treatment systems that can effectively reduce microbiological and chemical loads that are potentially present in source waters.

Treated water exiting the water-treatment plant contains a residual disinfectant to reduce potential microbiological risks posed by microorganisms inadvertently entering the distribution system. The dilution capacity of most source waters is enormous, so it would be extremely difficult to deliver sufficient levels of a harmful agent into a drinking-water treatment plant's raw water supply to cause harm.

On the other hand, contaminants can be introduced into a municipal drinking-water distribution system in sufficient quantities to cause problems. A large city contains thousands of miles of drinking-water pipes and multiple pumping stations. The potable water in these pipes provides continuous sources of water for domestic, industrial, and other uses. At critical junctions within the distribution system, valves and other control mechanisms have been installed to prevent the inadvertent backflow of non-potable water into the system. Intentional alteration of these mechanisms and the introduction of an agent could contaminate a portion of the distribution system.

Terrorists are not the only threat. In East Baton Rouge, Louisiana, investigators said an ammonia leak at a wastewater treatment facility was caused by criminals. They were trying to steal ammonia , which is a key ingredient in the manufacturing of illegal drugs such as methamphetamines.

Regarding physical security, members of Congress are considering whether protective measures should be focused on the largest water systems and facilities, where risks to the public are greatest, or on all systems and facilities, since small facilities are more vulnerable. Policymakers also are examining measures that could improve the coordination and exchange of information on vulnerabilities, risks, threats, and responses. This is a key objective of the Water ISAC and also of the DHS. This includes, for example, functions of the National Infrastructure Protection Center (NIPC) of the FBI, which brings together the private sector and government agencies at all levels to protect critical infrastructure, especially on cyber-security issues.

We need a greater focus on the vulnerability of the water-distribution system. It requires that an ongoing detection program be implemented to assure that any deterioration in water quality is quickly identified and followed by a rapid response and investigation. Some new equipment is being developed that will greatly speed the early detection of and testing for biological and chemical contamination. These devices include:

• DNA microchip arrays
• Immunologic techniques

- Microrobots
- Optical scan techniques
- Single-cell count and short-incubation dye tests

Past water-contamination situations show that risk has been elevated in nearly every case as a result of slow and inadequate monitoring, detection processes that often start after the event has occurred, and slow, poorly executed curative actions. Planning together in advance of an emergency can pay tremendous safety dividends.

Chapter 9

INFRASTRUCTURE SECURITY PROGRAMS

There are currently very few infrastructure security programs in place. Since 9/11, as I'm sure you expect, this situation has been improving. A number of larger utilities have instituted enhancements in physical security, but very few are improving cyber security, and there is very little effort in bio/chemical security. With the exception of some of the larger utility/energy companies, conducting basic background checks still remains a major issue.

The nation's natural gas utilities, for instance, have been examining their security, performing vulnerability assessments, and making appropriate enhancements to their security programs. These enhancements have included:

- Supplementing current emergency plans with terrorist risk elements
- Strengthening physical barriers
- Tightening control access
- Adjusting frequency of patrols

- Confirming response and recovery actions with local law and emergency officials

Within the American Gas Association (AGA), for example, 100 member companies have participated in committees focused on infrastructure security. These committees include Security, Integrity & Reliability Committee (Industry CEOs), Natural Gas Security Committee (physical), Operating Section Managing Committee, Technology Advisory Council (cyber), Operations Safety Regulatory Action Committee, Automation and Telecommunications (SCADA), Gas Control (Regional Planning), Supplemental Gas (LNG), Risk Management (Insurance), and Legal Committee. In the 1990s, one way the federal government attempted to address the potential threat to the U.S. information infrastructure is by providing a national perspective on the security-related challenges presented by the emergence of a National Information Infrastructure (NII).

At that time, a major national report found that electronic intruders are attacking data networks at increasing rates and had compromised elements of the telephone-signaling network. A senior Defense Information Systems Agency (DISA) official bluntly stated, "We are not prepared for an electronic version of Pearl Harbor," and he also said, "Our electronic infrastructure is not safe and secure." In one test of the security of Department of Defense (DoD) information systems, for example, a control team attacked nearly 10,000 systems using widely available techniques. They successfully penetrated 88 percent, of which only 4 percent were even detected. John M. McConnell, Director of the National Security Agency, said at the time, "We're more vulnerable than any other nation on earth."

This situation outside of DoD was no better. The telephone system, banking, credit, Federal Reserve systems, the stock exchanges, the power and fuels distribution systems, the air traffic control and other intelligent transportation systems, the federal elections system, public safety, and law enforcement all depended heavily then—and they depend even more heavily now—on networked information systems. All of these systems are potentially vulnerable to networked-based attacks. Informed estimates suggest that 90 to 95 percent of the information needed to carry out essential governmental functions must, in some way, be processed by information systems in the privately owned and operated parts of our critical infrastructures.

As former Department of Energy (DoE) Secretary Spencer Abraham left the department in 2005, he stated the need to improve the security of DoE facilities by:

- Enhancing cyber security programs to reduce the department's vulnerability to cyber threats and to improve classified data storage/use controls. Among the ways to accomplish this, he said, is to conduct more stringent probes and attacks on U.S. information systems through enhanced performance testing, including the use of no-notice vulnerability scanning and penetration testing.
- Consolidating sensitive national security assets to reduce the number of potential targets requiring enhanced protection
- Deploying contemporary advancements in security technologies to fortify DOE's "defense in depth" security posture and to implement more cost-effective security solutions
- Developing an "elite" protective force, analogous to the military's special operations forces, to secure DOE assets effectively against a wider spectrum of hostile threats
- Improving training programs and institutions in order to develop and maintain DOE's personnel within the security and safety disciplines

Secretary Abraham also discussed a consolidation program for special nuclear material within DoE sites around the United States, which he termed "by far the most effective measure that can be implemented to reduce the threat of a terrorist attack on the nuclear materials." In addition, the DOE is seeking to ensure that it has the highest quality and appropriate number of security forces personnel. This goal is instructive to leaders of our nation's utility and energy infrastructures.

Along with providing adequate numbers of its security force, the DOE is seeking to bolster its defense in depth strategy and to increase the availability of protective forces to train on critical tactical response skills. Secretary Abraham said, "As demonstrated by the nation's military forces in Afghanistan and Iraq, prudent and judicious application of technology can serve as a force multiplier to enhance protective force response to a hostile attack." This can be accomplished by:

- Increasing their tactical ability to bring effective fire upon an adversary force

- Enhancing their survivability against hostile fire
- Improving the command and control of responding forces

The implementation of security technology is intended to limit the DoE's dependency upon increased protective force staffing levels. The DoE has also instituted an "elite protective force" initiative to further enhance the knowledge, skills, and physical abilities of protective force personnel to effectively respond to a wider range of violent terrorist threats. Over the years the security police officers that comprise this protective force have maintained a set of challenging training qualifications and physical fitness standards similar to those of the best police organizations around the nation. DoE goes beyond the police-type standards to help protect the nation's nuclear weapons complex in response to the current terrorist threat.

In recent years, the DoE's protective force has transitioned from the classic police operational model to a more military-oriented model requiring an intensive focus on the move, shoot, and communicate skills normally attributed to those capabilities of elite military ground combat units. These units are modeled on the elite military units that defend our country, including the Special Forces, Rangers and SEALS. The DoE is working toward a time when all protective forces directly responsible for the protection of our critical infrastructures are transformed into such an elite fighting force. In order to accomplish this, the DoE is:

- Infusing protective force policy with an "expectation of excellence" and an emphasis on combating the armed terrorist threat
- Implementing medical and physical fitness standards necessary to support "elite force" performance capabilities
- Reorganizing protective force shifts into tactically cohesive units with appropriate operational command, control, and communications systems
- Reorienting protective force assignments to concentrate response resources in a tactical posture
- Requiring development of tactical response plans that locate protective forces either in direct defense of target locations or in appropriate ready response positions
- Relieving protective force personnel of non-essential routine duties to permit concentration on training and testing in support of their primary tactical mission

- Maximizing focus on tactical training by directing the initiation of training shifts in order for them to training as a team in the manner that they are expected to fight
- Instituting more rigorous and frequent performance testing to hone critical tactical skills in realistic and demanding environments
- Re-examining the applicability of current industrial safety standards against the levels needed to satisfy the intensity and rigor of required tactical training

EXPERIMENTAL PROGRAMS

The Department of Homeland Security (DHS) has four pilot programs designed to share threat information in ways that increase private sector security. Launched in cooperation with the FBI, these programs are running in Dallas, Seattle, Indianapolis, and Atlanta. Modeled after the FBI Dallas Emergency Response Network, the pilot programs expand the reach of DHS's Homeland Security Information Network (HSIN) initiative - a counter-terrorism communications tool that connects 50 states, five territories, Washington, D.C., and 50 major urban areas. The goal is to strengthen the exchange of threat information to critical infrastructure owners and operators in a variety of industries and locations, first responders, and local officials. As part of the HSIN-CI pilot program, more than 25,000 members of the network will have access to unclassified sector-specific information and alert notifications on a 24/7 basis.

When former Secretary of Homeland Security Tom Ridge began the first of the pilot programs, he said, "HSIN-CI connects our communities—the government community to the private sector community to the law enforcement community—the better we share information between our partners, the more quickly we are able to implement security measures where necessary." The HSIN-CI network allows local and regional areas to receive targeted alerts and notifications in real-time from DSH's Homeland Security Operations Center (HSOC) using standard communication devices including wired and wireless telephones, email, facsimile, and text pagers.

The network requires no additional hardware or software for federal, state, or local participants. The technical capacity of the network includes the ability to send 10,000 outbound voice calls per minute, 30,000 simultaneous inbound calls through an information "hotline," 5,000 simultaneous email messages, and 3,000 simultaneous facsimile transmissions in the event that information needs to be communicated.

In addition, HSIN-CI network, in partnership with the FBI, provides a reporting feature that allows the public to submit information about suspicious activities through the FBI Tips Program that is then shared with the Department's HSOC.

Idaho National Engineering and Environmental Laboratory (INEEL)

The DoE's Idaho National Engineering and Environmental Laboratory (INEEL) has mission testing systems that ultimately help electric utilities and system operators across the country protect their infrastructure, operations, and applications from physical and virtual attacks. INEEL technology mirrors real-world utility infrastructures.

A major control systems and emergency-management system vendor in the utility industry, ABB Ltd., is an early participant in INEEL's tests. ABB's software monitors and controls the flow of power transmissions and provides operators a view of transmission traffic. This vendor recently paid TecSys Development Inc. to include ABB's new emergency-management software, Network Manager, in TecSys' roster of supported products. TecSys makes the ConsoleWorks Intelligent Event Manager central-monitoring technology.

ConsoleWorks gives customers a single view of systems. It also provides alarms and suggested processes to avert downtime and potential security problems. Traditionally, this solution supports many IT systems including Cisco routers, HP's HP-UX operating system, and Oracle's database-management software. This is the first time it has supported an emergency-management system for the utility industry. Utility operators who use Network Manager and run ConsoleWorks can get a single view of both IT and operational systems. This enhances their ability to monitor possible cyber intrusions and other problems. The bundled offering has important implications for the industry.

One ConsoleWorks customer in the utility industry says that the software lets him find the root cause for an IT system outage and resolve it more quickly. Problems that remained unsolved for weeks are now closed within half an hour.

NERC Standards

Traditionally, NERC has published a standard for securing utility operators' control centers' cyber security, including such best-practice requirements as enacting procedures for firewall management, secure dial-up modem connections, and antivirus software. The new standards, which are summarized in Chapter 3, are for substations, and they include best practice requirements. It's important, however, for utilities and system operators to first identify which substations are most critical. It will take years to implement cyber security procedures for the thousands of substations.

According to Lou Leffler, manager of critical infrastructure protection at NERC, substations represent a major problem to locking down cyber security for utilities. "Substations have hundreds of times the connections that most anti-virus software is accustomed to dealing with, and their real-time operating systems mean they can't tolerate a millisecond interruption. What's more, many of them were built for a specific task during a time when we didn't have cyber security concerns." Bandwidth and timing issues necessary for add-on security systems represent a major challenge.

But everyone I've talked with realizes the need to harden cyber security defenses. Consultant Joe Weiss says, "The highest probability of cyber security problems comes from viruses and worms. At a utility, next comes corporate IT personnel installing the latest version of antivirus software. Loading the latest version of antivirus software on HP-UX or Solaris could slow down the operation, thus shutting down multiple real-time control systems."

Weiss can point to several outages that have occurred because of cyber security breaches, including breaches of servers, which are often attached to components that are linked to the grid. And, he adds, "That's just the tip of the iceberg."

GLOBAL EFFORTS AND COORDINATION

There are a great many global efforts to coordinate increased security at utility, energy and other critical infrastructures around the world.

The World Summit on the Information Society

One of them, the World Summit on the Information Society (WSIS), recognizes the importance of a commitment from the private sector in developing and diffusing information and communication technologies (ICTs) for infrastructure, content and applications. WSIS has established targets, which consider different national circumstances. The targets include the following goals:

- Connect villages with ICTs and establish community access points
- Connect universities, colleges, secondary schools, and primary schools with ICTs
- Connect scientific and research centers with ICTs
- Connect public libraries, cultural centers, museums, post offices, and archives with ICTs
- Connect health centers and hospitals with ICTs
- Connect all local and central government departments and establish websites and email addresses
- Adapt all primary and secondary school curricula to meet the challenges of the Information Society, taking into account national circumstances
- Ensure that all of the world's population has access to television and radio services.
- Encourage the development of content and to put in place technical conditions in order to facilitate the presence and use of all world languages on the Internet
- Ensure that more than half the world's inhabitants have access to ICTs within their reach

Each country is encouraged to establish at least one functioning Public/Private Partnership (PPP) or Multi-Sector Partnership (MSP) as a showcase for future action. Each country is also charged with identifying mechanisms, at the national, regional, and international levels, for the initiation and promotion of partnerships among stakeholders of the Information Society. ICTs allow people anywhere in the world to access information and knowledge almost instantaneously. The goal is for individuals, organizations and communities to benefit from increased access to knowledge and information.

An Australian Example

Australia has introduced new legislation to protect against terrorism. As a result, specific compliance requirements must be met, demanding that operators provide measures to both minimize the likelihood of terrorist incidents and mitigate the resultant damage after an event. Under the terms of the National Counter Terrorism Agreement, the State of Victoria, which includes Melbourne, has identified and developed a database of more than 600 components of critical infrastructure. The operators of these assets, considered Essential Service Providers, must develop risk a management plan. The main objectives of this plan are to:

- Prevent terrorist acts in relation to the declared essential service
- Mitigate the effects of a terrorist act.
- Recover the service from a terrorist act, and to ensure continuity of the service at all times

Now, managers must examine and consider the nature and the likelihood of a terrorist incident either in their facility or close by, and they are to take steps to defend against the risks and consequences of these potential attacks. If they don't, their company is exposed to legal liability. Worse yet, management can be considered culpable. Employers can also be considered negligent if they do not take reasonable steps to eliminate or diminish known or reasonably foreseeable risks that could cause harm.

The thrust of this Australian plan is to place the onus with the private operator and adapt an overseeing role for the assigned government agency to ensure that the appropriate best practice is met and that benchmark security standards are reached. Part of the challenge is for critical infrastructure security managers to merge the organizational and functional elements for both IT security and physical security. As is well known in the United States, these two areas are often at different poles within the business process. Knowledge gaps often exist that allow inherent weaknesses to flourish within the company.

So some Australian organizations have appointed one person to speak for both IT security and physical security areas. Then that person establishes lines of communication directly to the upper echelon of corporate management and seeks the support of senior executives. "Security is about educating the business leaders about the threats the organization faces, the likely negative consequences and costs of those threats, and the necessary control measures that need to be implemented as effective safeguards," says Allen Fleckner, Director of Security and Risk, Emergency Management Experts, Melbourne. "To raise the security culture of the organization, it is important to develop an omnipresent security program that promotes effective security as an essential business reality. This approach avoids a weaker security stance based on the dangerously outdated notion that it "won't happen to us," and improves lines of communication directly through to the executive level. In turn, it coordinates an approach to integrate physical security, information protection, and risk management."

MONITORING CAPABILITIES

For the DHS to monitor all utility/energy security issues it would require using a satellite and having security notification capabilities built into all systems used by the utility/energy industry. The federal government and private industry, however, have done very little to secure our satellite systems. If you want to find the most-ignored cyber security hole in America's critical infrastructure, some Congressional investigators say, "Look up!"

Critical commercial satellite systems relied upon by federal agencies, civilians and the Pentagon are potentially vulnerable to a variety of sophisticated hack attacks that could cause service disruptions. They could also send a satellite spinning out of control. A recent report from the General Accounting Office (GAO) criticized the White House for not taking the vulnerabilities into account in its national cyber security planning. This frequent criticism extends back to the Clinton administration.

The focus of the report is on satellite systems that are used extensively by the federal government but, as is the case with most of our nation's critical infrastructures, are controlled by the private sector. Among the weaknesses GAO investigators found are the following:

- Some satellite companies don't encrypt the tracking and control uplinks through which the satellites are controlled from the ground, making them vulnerable to spoofing, "which could cause the spacecraft to destroy itself," the GAO reported.
- "It is also feasible to insert false information or computer viruses into the terrestrial computer networks associated with a space system, either remotely or through an on-site connection," the GAO found. "Such an attack could lead to space system degradation or even complete loss of spacecraft utility."
- An attack could impact military operations, the report claimed. It's important to note that commercial satellites were used for 45 percent of all communications between the U.S. and forces in the Persian Gulf region during Desert Storm.
- The study also found that there are some significant safeguards in place. In certain cases, for example, companies deliberately use extremely high-power transmitters to control their satellites. This approach makes it less likely that an attacker could overpower the authentic signal with a fake one.

- But the level of security varies significantly, the report found, and with little regulation governing satellite security, commercial providers have little incentive to invest in costly solutions.

A federal policy initiated in 2001 requires satellite providers handling national security communications to meet minimal cyber security standards. But the GAO report found that not a single company was entirely compliant with the directive, which has no enforcement mechanism. "At times, noncompliant satellites have been DOD's only option," the report stated. Except for GPS vulnerabilities, the report found that satellite systems were ignored in President Clinton's cyber security efforts. Unfortunately, they are faring no better under the Bush administration's cyber security push.

"Given the importance of satellites to the national economy, the federal government's growing reliance on them, and the many threats that face them, failure to explicitly include satellites in the national approach to [critical infrastructure protection] leaves a critical aspect of the national infrastructure without focused attention," the GAO concludes.

Chapter 10

FINANCING THE UTILITY INDUSTRY SECURITY

For the great majority of American companies, the three Gs—guards, gates, and guns—continue to be their main security strategies. Others have extended their security budgets for capital equipment and additional personnel. In the latter group, many have:

- Conducted assessment reviews
- Installed technology safeguards
- Improved former vulnerabilities, such as business continuity as it relates to supply chains
- Placed physical and IT security under one corporate officer, with this person often reporting directly to the CFO

As I've said throughout this book, companies in the utility and energy industries are likely to be special targets of terrorists. Here's some excellent advice from a security expert I've known for years. "It's important for companies to do the research so they can learn what is state-of-the-art," says Joe

Gulienello of Securitas Security Services USA. "Compare products, not only from vendors but also from your peers. What are companies comparable to yours using to increase their security? Find out by asking them."

Some security executives have trouble requesting additional funding because it's hard to measure the return on investment for improvements related to the possibility of terrorism. Joe Gulinello has good advice here, too. "Do some benchmarking. Survey other companies in your region. Include companies similar to yours and those that provide different types of services, too. Learn what they are doing, what you think your utility or energy company should do, then present your argument to management."

North Carolina-based Duke Energy has given special attention to calculating its security costs. Although I know of no specific return-on-investment (ROI) financials that have been prepared, Duke Energy has developed cost-effective solutions for future threats. They have, for example, licensed what they consider the best software to help with the process, not the most expensive or the cheapest. Instead, Duke Energy selected an off-the-shelf product that met Department of Energy (DOE) requirements and was relatively easy to use.

This company has also quantified some of its security costs, including, for example, the cost of going from a yellow to an orange alert. Here's what they found: For an office complex that employs 2,000 people in three buildings, the additional operating cost for contract labor and extra hours of coverage is $10,000 per week.

One rule of thumb I've heard within the security industry is to allocate security funding this way:

- 40 percent to security personnel
- 40 percent to processes, such as systems to handle crisis-management hotlines
- 20 percent to technology, including computers, lighting, and alarm systems

Whatever increased security costs have been incurred or are planned, utility and energy executives want to know how they will at least get their money back.

RATE INCREASES

One of the most common ways to get reimbursed for these additional expenses is through rate increases. Shortly after 9/11, the Federal Energy

Regulatory Commission (FERC) told companies that it would approve costs to upgrade security. This commission, which regulates rates for wholesale electricity and natural-gas shipments, defined the expenses as "prudently incurred costs necessary to further safeguard the reliability and security of our energy and supply infrastructure."

Public-service commissioners around the country that I've talked with want to be sure critical infrastructures in their states are protected. They want utility and energy companies to take "necessary" actions with "reasonable" costs that will become the basis for modest rate increases. State regulatory bodies oversee and set prices for most utilities and decide whether to grant rate increases through a series of hearings.

There's no clear-cut way to determine exactly how much the cost of additional security adds to a customer's gas, electric, or water bill. There are too many variables within different utility and energy companies, which can also vary from region to region. In addition, enhanced security can extend over a wide range of actions, including:

- Installing better locks
- Using heavier chains
- Employing sophisticated motion-detection devices
- Installing complex cyber-security systems
- Engaging in new construction that hardens portions of utility facilities and infrastructure

Here are a few of the many cases of utility companies winning approval of rate increases for added security costs. In Pennsylvania, American Water won a $30 million rate increase that included some incremental security costs attributed to 9/11. The typical monthly residential water bill for a customer in Pennsylvania at the time was $38 per month. Of that amount, about 92 cents could be traced to the cost of increased security.

In Michigan, Consumers Energy Company won approval for higher rates to pay for $24.5 million in security measures at its nuclear plant. These improvements were required under federal guidelines.

In Iowa, utilities won rate increases for $1.6 million in security costs. For example, MidAmerican Energy, which serves 460,000 people, won a gas rate case that included $780,375 in security costs. It added only 27 cents per year to consumers' bills.

The National Association of Regulatory Utility Commissioners (NARUC) issued a white paper on the subject that states that, "state commissioners are also sensitive to not creating an incentive for carte blanche expenditures on security, which in turn end up directly in higher rates." The research arm of the NARUC, the National Regulatory Research Institute, issued a study. It found that utilities in 22 states have filed to recover security costs. Utilities in another 11 states are aware of security costs but have not yet filed to recover them. Some utilities have put off security investments until state regulators assure the companies that they can reflect the costs of these investments in adjusted rates.

GOVERNMENT INCENTIVES

While we continue to wage a war against terrorists in Iraq, Afghanistan, and elsewhere around the world, we are not, however, doing nearly enough to protect our homeland. The federal government is providing minimal incentives, primarily approving rate increases. That's it.

One of the major consequences of this lack of incentives is that private enterprise is providing minimal security. On the whole, companies are protecting themselves in ways that are similar to their competition, no more, no less. In a survey conducted by the Council on Competitiveness only one year after 9/11, more than 90 percent of executives did not believe terrorists would target their companies, and barely 50 percent said that their companies had increased their security spending between 2001 and 2002. More recent surveys have shown that increased security continues not to be a high priority.

Please notice that I am not focusing on the utility and energy industries but, instead, am discussing all types of companies in the United States. Increased security will not happen by itself. It costs money. If particular companies get far ahead of competitors in making their facilities and employees more secure, their costs of doing business are higher. So it's harder for them to compete.

Stephen E. Flynn is a Senior Fellow in Natural Security Studies at the Council on Foreign Relations. He has written the following in Foreign Affairs magazine: "The invisible hand of the free market simply will not provide sufficient economic incentives for private companies to protect from acts of terrorism the global networks that they largely own and operate. This is because their executives worry that such investments would place them at a competitive disadvantage." If the government provided adequate incentives, two goals would be accomplished:

- Companies, particularly utility and energy firms with their critical infrastructures, would have some of the funding necessary to increase their security.

- Competitors would realize that others within their industries were raising the security bar uniformly, so no one company would have a major competitive advantage.

Another type of government incentive is to require security upgrades according to standards that every company in an industry must meet. Without government funding, that strategy causes strain on companies and eventually increased costs that are most likely passed on to customers.

A very different approach to protecting our critical infrastructures is involving the private sector in developing security standards and creating new mechanisms to enforce them. Randolph Lerner, Chairman of MBNA Corporation, a leading bank holding company, suggests that the structure of the Federal Reserve System might provide the framework for a far more secure homeland-security program. This new model would be called the Federal Security Reserve System (FSRS) and would include self-funding mechanisms to engage American companies in protecting our critical infrastructures from the major to devastating disruptions that result from terrorist attacks.

The FSRS, among other acts, would create incentives for expanded investment in security through a mandatory insurance program. Utilities, energy companies, and other owners of our critical infrastructure, including healthcare, water, food, and financial sectors, would be required to carry defined levels of insurance. The insurance requirement would lower the reliance on public resources after acts of terrorism. The insurance requirement would also involve the insurance industry as a vital partner with the owners of our essential infrastructures.

The FSRS would have a national board of governors, ten regional homeland-security districts, and 92 local branches. Members of the board would work closely with the Department of Homeland Security (DHS), testify before Congress, interact with the National Academy of Sciences, and communicate with international security groups. Each of the ten homeland-security districts would be responsible for a particular sector. Texas, for example, could be the lead in oil and gas security; the Northeast would spearhead financial security.

The overriding design of the FSRS is to create an organization that does not rely on government agencies for existence, funding, or direction. The fate of the nation is tied to our critical infrastructures, which are primarily in the control of private industry. Instead, the FSRS would create a structure that involves participation of companies and individuals in new ways that are likely to make our nation safer.

INSURANCE COMPANY INCENTIVES

"The Terrorism Risk Insurance Act (TRIA) is law," says Robert Hartwig, Chief Economist with the Insurance Information Institute in Washington, D.C. "So as long as Congress extends this act, this insurance option will continue to be offered." Mr. Hartwig points out that the price of this insurance is "a function of the perception of risk, so different types of facilities get different rates." Among the variables are the location of the facility, the size and value of the physical plant, the amount of intellectual property (IP), and calculations for business interruption and worker's compensation losses.

Under TRIA, the Treasury Department covers 90 percent of terrorism claims after an insurer's exposure exceeds 15 percent of its commercial premiums in 2005. The bill states that federal assistance would not have to be repaid if total industry terrorism claims are more than $10 billion in $15 billion in 2005. Below this threshold, insurers would have to reimburse the Treasury Department for its aid by imposing surcharges of up to 3 percent on their commercial premiums. The law also consolidates civil lawsuits stemming from a terrorist attack in a single federal court for trial under the laws of the state in which the attack took place.

TRIA, however, is not a federal-government backstop to insurance losses caused by terrorism. And there is no widely accepted model for assessing the risk of terrorism. Some third-party terrorism policies can be found, as can some specialty lines, including:

- The Nuclear Electric Insurance Limited's (NEIL) coverage for nuclear-power-plant property damage
- Oil Insurance Limited's (OIL) petroleum-industry product
- AEGIS offers terrorism coverage for utility companies and AEGIS members,
- The Air Transport Association offers a terrorism insurance product that they distribute through Equitime, which is their captive insurer creation.

In Europe, special partnerships address terrorism coverage shortages. Special Risk Insurance and Reinsurance Luxembourg (SRIR), for instance, has been established by Zurich Financial Services, XL Capital Ltd., Swiss Re, SCOR, Hannover Re, and Allianz. SRIR has reported a total committed capital of EUR 500 million. Note that its policies only cover damage to property directly from an act of terrorism and will only be focused on Europe. Business-interruption and liability losses are not insured.

In the United States, the market is served by six insurance companies so there is some availability. But can a large utility or energy company buy enough at a reasonable price? To some extent, the answer depends on the limits you will accept. The National Association of Insurance Commissioners (NAIC) has agreed to rethink its plan for mixing catastrophe reserving with terrorism risk. This action occurred after the National Association of Mutual Insurance Companies (NAMIC) and other trade organizations cautioned against combining the two. Some regulators disagree. They suggest that adding terrorism risk to catastrophe reserving would make the plan more viable and would entice Congress to grant terrorism deductibility for the whole package.

Meanwhile, many people in the insurance industry consider terrorism risk not insurable because it can't be quantified. That's creating a market for companies to develop new modeling tools that analyze the risk of terrorism. Others believe it will take a few more years before the insurance industry will develop the means to cover terrorism by use of approaches that are similar to those that handle the risks of natural catastrophes. The federal government is likely to be called upon to issue guidelines and to serve as an insurer of last resort. Congress, however, is not reacting much at all.

In a survey by the Building Owners and Managers Association International (BOMA), which lobbied in favor of TRIA for more than a year, 27 percent of respondents said they were unable to obtain terrorism insurance at any cost. Of the 73 percent who said they could obtain coverage, 80 percent said they encountered some type of restriction and/or a premium increase of between 20 and 200 percent.

Mortgage bankers, developers, and some construction labor unions also pushed for the legislation. They said that bond debt ratings for commercial properties had suffered and building projects were held up due to scant terrorism insurance. Insurers said they could absorb the billions in claims from the 9/11 attacks but would be ruined by another similar event.

RESELLING OF UTILITY SECURITY SERVICES

I believe utilities can offset the cost of providing vastly improved security for themselves by reselling security services to their key industrial and commercial accounts. This concept is discussed at length in Chapter 12, but I want to begin here with a model provided by the healthcare industry and then introduce some of the services that the utility/energy industry might offer its customers.

Healthcare Industry

All healthcare facilities are supposed to have adopted the following policies and procedures, which must be documented:

- Training programs in security-management and process issues
- Formal data-processing protocols
- Formal protocols for controlling access to data
- Internal audit procedures
- Certification of data systems for compliance with Department of Health and Human Services (DHHS) security standards
- Chain-of-trust agreements with covered entities with which healthcare organizations exchange electronic information
- Contingency plans that ensure continuity and preservation of data in the event of an emergency
- Security features for initial clearance of all personnel who have access to health information along with ongoing supervision, training, and monitoring of these personnel
- Security-configuration management procedures, including virus checking, hardware and software systems review, and documentation
- Specific procedures when personnel terminate employment
- Security-management structure that maintains continual risk assessment and sanction policies and procedures

Physical Safeguards

Data and data systems must be physically protected from intrusion and environmental hazards in the following ways:

- Designation of a specific person for responsibility of security
- Controlling access to and altering of computer hardware
- Enforcement of "need to know" clearances
- Implementation of workstation security activities
- Development of disaster/intrusion response and recovery plans

- Maintenance of security records
- Implementation of identity-verification procedures for personnel in order to physically access sites

Technical Security Services

Software control and procedures regarding stored data include these requirements:

- Providing for internal audits and controls within data systems
- Control access by users through authentication
- Ensure that stored data is neither altered nor inappropriately accessed/processed
- Allow data access to particular privileged classes of personnel, including during crises

Technical Security Mechanisms

To ensure that data cannot easily be accessed, intercepted, or interpreted by unauthorized third parties, the following requirements focus on accessed data and the transmission of stored data:

- Validation that stored data being transmitted is accurate
- Validation that received data is identical to sent data
- Data transmissions either encrypted or controlled by a dedicated, secure line. If transmissions are not encrypted, DHHS would also require three elements:
  - Alarms to signal abnormal communication conditions
  - Automatic recording of audit trail-information
  - A method for authentication of the entity receiving the data

Schools and Colleges

Utility/energy companies can consider offering security services to a wide range of educational institutions. If you do, of course you'll find that school security is a top priority for parents as well as government and school officials. School systems across the country rely on the latest high-technology systems and tools to protect children. These devices, from a list supplied by officials in New Mexico, include:

- High-resolution cameras
- Motion detectors
- Stationary and hand-held metal detectors
- Bar-coded student and staff ID cards

In New Jersey, state and school officials have implemented a number of steps to make schools safer. These include:

- A comprehensive school-security checklist that includes specific guidelines to ensure school safety
- Security audits of each school. Security experts visit every New Jersey school to review their enhanced safety.
- Security courses for school personnel. Department of Homeland Security officials have helped establish classes that teach school personnel how to recognize and react to potential terrorist activity.
- Inspection of school construction and maintenance activities. Law-enforcement personnel visit every school construction site to help ensure that security systems are in construction plans.
- A school-security summit. To bring the best ideas from New Jersey and the nation together in one place, Rutgers University has hosted a forum on school security.

New initiatives used in schools can be extended to protect other places where people gather, including malls and train and bus stations. These extended initiatives can include:

- Smart security cameras that can be programmed to recognize specified suspicious behaviors
- These smart cameras also can share videos from the site directly with local police. First responders can then know what is happening inside a building before they arrive.

As you can see, the utility/energy industry is caught between a rock and a hard place. If executives in this industry upgrade their security systems, it makes it difficult for them to compete in an open market. But if they don't upgrade security, they are at high risk if a terrorist attack were to occur. Because we are at war and our economy and national security totally rely on the utility/energy industry, I believe we need to move now at the national level to secure these vital infrastructures, and it needs to be a national effort to ensure fairness to all.

Chapter 11

ROLE OF NATIONAL AND STATE ASSOCIATIONS

NATIONAL UTILITY ASSOCIATIONS

Many national utility/energy associations can provide information and shared experience to managers at all levels. This chapter provides introductions to the major national utility/energy associations and a brief description of the objectives and positions they have taken on national-security issues and terrorism. These associations play a vital role in promoting the protection of our nation's utilities through lobbying efforts, training, financing, and collaboration among smaller utilities.

Edison Electric Institute

Edison Electric Institute (EEI) is the leading trade association for U.S. shareholder-owned electric companies. EEI also serves international affiliates and industry associates worldwide. U.S. members provide services to more than 90 percent of the ultimate customers in the shareholder-owned segment of the industry and nearly 70 percent of all electric-utility ultimate customers in the nation. U.S. member utilities generate almost 70 percent of the electricity produced.

EEI provides advocacy, authoritative analysis, and industry data to its members, Congress, government agencies, the financial community, and others. The forums of this organization enable member-company representatives to discuss issues and strategies that advance the industry and help to ensure a competitive position in the marketplace. EEI presents its mission as ensuring members' success by:

- Advocating public policy
- Expanding market opportunities
- Providing strategic business information

North American Electric Reliability Council

Since 9/11, the North American Electric Reliability Council (NERC) has worked to help ensure the protection of the nation's critical infrastructures. Among NERC's actions are the following:

- Coordinating its threat levels according to national levels
- Implementing twelve security guidelines, with two2 more being developed
- Developing a cyber-security "standard" for the industry

Electrical generators make use of coal, nuclear power, natural gas, fuel oil, hydroelectric dams, and other renewable sources of energy. The mix of fuels used to generate electricity varies from region to region, yet on a national average, little of our electricity is generated using imported oil. According to NERC, "This country's electrical supply is less susceptible to disruption or supply problems than would be the case if we relied on a single fuel source."

The Nuclear Energy Institute

The Nuclear Energy Institute (NEI) maintains that its current security measures at nuclear-power facilities are highly effective in deterring sabotage and safeguarding the public. The FBI considers nuclear-power facilities "hardened targets" due to their security measures and "defense-in-depth" strategy.

Nuclear power plant containment buildings, which house the reactors, are extremely robust. The containment buildings, which are made of steel and concrete materials up to four feet thick, have been designed to withstand the impact of hurricanes, tornados, and floods. The reactors themselves are contained in steel shells that are about six inches thick. Although specific design requirements vary at each facility, containment buildings are designed to withstand airliner accidents. In addition to the comprehensive safety measures already in place, the industry works with local, state, and federal officials to explore ways that nuclear-plant security can be further enhanced.

American Gas Association

The American Gas Association (AGA) advocates the interests of its energy-utility members and their customers. AGA also provides information and services promoting demand and supply growth and operational excellence designed to result in the safe, reliable, and cost-competitive delivery of natural gas. To further this mission, the AGA:

- Focuses on the advocacy of natural-gas issues that are priorities for the energy membership and that are achievable in a cost-effective way
- Encourages and assists members in sharing information designed to achieve operational excellence by improving their security, safety, reliability, efficiency, and environmental and other performance
- Assists members in managing industry transition and in responding to customer energy needs, natural-gas markets, and emerging technologies
- Collects and analyzes timely data and disseminates information to policy makers and the public about energy utilities and the natural-gas industry
- Serves as a voice on behalf of the energy-utility industry promoting natural-gas demand by emphasizing, before a variety of audiences, the energy efficiency, environmental, and other benefits of natural gas and promoting natural-gas supply by emphasizing public policies favorable to increased supplies and lower prices to consumers

Electric Power Research Institute

The Electric Power Research Institute (EPRI) was established to address many critical issues. According to EPRI, its founding occurred as a result of:

- The great northeastern blackout of 1965 in the United States, which revealed for the first time serious vulnerabilities of the nation's electricity-supply system
- War in the Mideast in 1967, which foreshadowed the energy crises of the 1970s and the danger inherent in our growing dependence on foreign oil
- The National Environmental Policy Act of 1969, which launched decades of environmental legislation that would challenge utilities' scientific and technical capabilities

The convergence of these issues had many questioning the ability of the existing energy infrastructure to effectively serve the needs of the nation. According to many authorities, utilities had too little invested in the science and technology required to adapt to changing needs. And utilities did not have the resources or the expertise to deal with the technical issues. In 1971, threatened by Senate proposals to create a federal agency to conduct electricity-related R&D, America's public and private utilities banded together to develop an industry-organized alternative, the Electric Power Research Institute (EPRI).

EPRI began operations in 1973. It took over R&D projects previously managed by others and quickly expanded to address many of the industry's most important issues. According to EPRI, its scientific and technical achievements have returned benefits to utilities and their customers estimated at more than $50 billion—nearly 10 times the collective utility investment in EPRI.

In one of its most important programs, EPRI formed the Infrastructure Security Initiative (ISI). It is designed as a tightly focused effort to identify key vulnerabilities in the electric-power grid and to design immediately applicable countermeasures. ISI grew out of EPRI's response to the terrorist attacks of 9/11. Building on experience gained from the western states' power crisis, EPRI assembled its own interdisciplinary team to prepare an EPRI technical report, "Electricity Infrastructure Security Assessment." This report provided a preliminary analysis of potential terrorist threats, and ISI was established to put the recommendations of this report into action. ISI explores the following areas:

- Recovery transformers
- Vulnerability assessments
- Secure communications

Figure 11.1 Working to keep reliable power flowing to consumers

- "Red team" attacks
- Immediate countermeasures

After a terrorist attack, a critical element in enabling the recovery of the U.S. electric grid is to have recovery transformers designed for rapid deployment and installation. This strategy is based on storing this equipment at strategic U.S. locations. Transformers typically need to be customized for specific uses, but they take about 1½ years to build, transport, and install. ISI is developing new transformers with multiple MVA ratings for connection at multiple voltages and sized to allow for transportation by rail, truck, and/or large cargo planes.

Among the main tasks of the ISI planning phase was to develop conceptual designs for recovery transformers that would have an expected lifetime of at least two to three years and also enable the rapid replacement of damaged transformers. Two major transformer manufacturers are working on designs for the recovery transformers. The resulting work is based on existing technology. Installation time for a recovery transformer, once transported, is estimated to be two days.

During ISI phase 1, the two manufacturers will complete the design for multiple voltage/MVA classes of recovery transformers; then they will build and factory-test two to four prototype units at different specified sizes. During phase 2, the prototypes will be delivered to host utilities for rapid deployment and field testing. The overall goal of this work is to develop a vulnerability-assessment methodology capable of determining the impact of potential terrorist attacks anywhere in the electricity-supply chain. During the planning phase, a new procedure was developed to identify and rank critical simultaneous multi-station contingencies, which might be expected from a coordinated terrorist attack.

In ISI phase 1, the new procedure will be developed into a "Self-Assessment Guidebook" for qualitative identification and ranking of simultaneous contingencies. These include dependencies on natural gas and other critical infrastructures. In addition, ISI members will be given a software tool so they can quantitatively evaluate the vulnerabilities of their own power systems.

EPRI is also focusing substantial efforts to make cyber-security systems more secure. ISI work in the planning phase, for example, concentrated on assessing communication-system vulnerabilities that produced EPRI Technical Report 1008988, "Scoping Study on Security Processes and Impacts." Utilities are using this report to identify communication-system vulnerabilities and evaluate alternative countermeasures for enhanced security.

Phase 1 will provide specific measures to enhance communication security in three specific areas:

- Develop procedures to secure the transfer of data based on the common information model (CIM) and for protection of the information during and after its transmittal to a server
- Design a monitoring and mitigation system to protect utility supervisory control and data-analysis devices from malicious or anomalous commands
- Develop and demonstrate ways to enhance utility communication networks/systems

To demonstrate best practices for computer security, EPRI is among many organizations that probe for weaknesses in computer and information networks by launching mock assaults using a "red team" effort. During the ISI planning phase, for example, probes were made of the cyber systems of two volunteer host utilities. Once the assault and analyses were complete, each volunteer utility was provided a confidential documentation of the results. A generic report was then distributed to ISI funders. During the planning phase, red-team attacks concentrated on Internet-connected corporate IT systems. The lessons learned from these efforts included the following:

- Formation of an effective incident-response team at utilities. This team represents an important way to assess and mitigate impacts of future cyber threats
- Numerous security "weak links" that include remote locations and business partners tied to a utility's cyber network
- Periodic employee training, which is key to an effective security program

During phase 1, ISI is performing mock assaults on volunteer host utility systems directly through hardware (e.g., remote terminal units (RTUs) or through communication systems (e.g., microwave links). This work is expected to provide a realistic assessment of utility vulnerabilities from communication systems.

Figure 11.2 During an ice storm, crews brave dangerous conditions.

United Telecommunications Council

The United Telecommunications Council (UTC) considers itself a global trade association dedicated to creating a favorable business, regulatory, and technological environment for companies that own, manage, or provide critical telecommunications systems in support of their core business. Founded in 1948 to advocate for the allocation of an additional radio spectrum for power utilities, UTC has evolved into an organization that represents electric, gas, and water utilities; natural gas pipelines; critical infrastructure companies; and other industry stakeholders. From its headquarters in downtown Washington, DC, UTC provides information, products, and services to help members:

- Manage their telecommunications and information technology more effectively and efficiently
- Voice their concerns to legislators and regulators
- Identify and capitalize on opportunities linked to deregulation worldwide
- Network with other telecom and IT professionals

UTC is also an authorized certified frequency coordinator for private land mobile radio services below 512 MHz and between 800 and 900 MHz. In addition, UTC is the sole frequency coordinator authorized to coordinate channels previously allocated exclusively to the power radio service. UTC also maintains the national power line carrier (PLC) database for the coordination of PLC use with licensed government radio services in the 10 to 490 kHz band.

National Rural Electric Cooperative Association

National Rural Electric Cooperative Association (NRECA) is "the national service organization dedicated to representing the national interests of cooperative electric utilities and the consumers they serve." The NRECA board of directors oversees the association's activities and consists of 47 members, one from each state in which there is an electric distribution cooperative. Founded in 1942, NRECA was organized specifically to overcome World War II shortages of electric construction materials, to obtain insurance coverage for newly constructed rural electric cooperatives, and to mitigate wholesale power problems. NRECA continues to be an advocate for consumer-owned cooperatives on energy and operational issues as well as rural community and economic development.

NRECA's more than 900 member cooperatives serve 37 million people in 47 states.

Most of the 865 distribution systems are consumer-owned cooperatives, and some are public power districts. NRECA membership includes:

- Other organizations formed by these local utilities
- Generation and transmission cooperatives for power supply
- Statewide and regional trade and service associations
- Supply and manufacturing cooperatives
- Data-processing cooperatives
- Employee credit unions

Associate membership is open to equipment manufacturers and distributors, wholesalers, consultants, and other entities that do business with members of the electric cooperative network.

The National Rural Utilities Cooperative Finance Corporation

The National Rural Utilities Cooperative Finance Corporation (CFC) offers full-service financing, investment, and related services to its members. CFC is a private market lender for the nation's electric cooperative network. This association seeks to ensure that its lending and investment programs—as well as all related financial and business- management services—deliver high quality at a reasonable cost. CFC provides members with competitively priced financing through its role as a conduit to the domestic and international capital markets. CFC raises funds for its loan programs with the support of its members' equity and investments and through the sale of multiple financing vehicles in the domestic and international financial markets. CFC is an independent source of financing that supplements the credit programs of the U.S. Department of Agriculture's Rural Utilities Service (RUS). It also provides financing and business assistance to non-RUS borrowers.

North American Electric Reliability Council

The North American Electric Reliability Council (NERC) is a not-for-profit corporation whose members consist of ten regional reliability councils.

The members of these councils come from all segments of the electric industry:

- Investor-owned utilities
- Federal power agencies
- Rural electric cooperatives
- State, municipal, and provincial utilities
- Independent power producers
- Power marketers
- End-use customers

These segments account for virtually all the electricity supplied and used in the United States, Canada, and a portion of Baja California Norte, Mexico. NERC states that its mission is to ensure that the bulk electric system in North America is reliable, adequate, and secure. Formed in 1968, NERC is a voluntary organization relying on reciprocity, peer pressure, and the mutual self-interest of all those involved. Through this approach, NERC has helped to make the North American bulk electric system among the most reliable in the world. To fulfill its mission, NERC:

- Sets standards for the reliable operation and planning of the bulk electric system
- Monitors, assesses, and enforces compliance with standards for bulk electric-system reliability.
- Provides education and training resources to promote bulk electric-system reliability
- Assesses, analyzes, and reports on bulk electric-system adequacy and performance.
- Coordinates with regional reliability councils and other organizations
- Coordinates the provision of applications, data, and services that are necessary to support the reliable operation and planning of the bulk electric system
- Certifies reliability service organizations and personnel
- Coordinates critical infrastructure protection of the bulk electric system
- Enables the reliable operation of the interconnected bulk electric system by facilitating information exchange and coordination among reliability service organizations

- Administers procedures for appeals and conflict resolution for relia-bility-standards development, certification, compliance, and other matters related to bulk electric-system reliability

NERC believes that the *voluntary* system of compliance with reliability standards is no longer adequate. So NERC is promoting the development of a new *mandatory* system of reliability standards and compliance, back-stopped in the United States by the Federal Energy Regulatory Commission. NERC plays a major role in protecting the electric system by serving as the focal point for coordinating information exchange on critical infrastructure issues between the electricity industry and the federal government. Through NERC, the government and industry work together to protect the electricity infrastructure from physical and cyber attacks. This coordination helps to ensure that the industry speaks with one voice and takes consistent and effective action.

The Department of Energy (DOE) has designated NERC as the elec-tricity sector coordinator for critical infrastructure protection. The National Infrastructure Protection Center (NIPC) has asked NERC to be the information sharing and analysis center for the electricity sector. NERC also works closely with the Department of Homeland Security (DHS) to ensure that the critical infrastructure-protection functions, so vital to the industry, are fully integrated and coordinated with DHS.

American Public Power Association

The American Public Power Association (APPA) is the service organiza-tion for the nation's more than 2,000 community-owned electric utilities, which serve more than 43 million Americans. Created in 1940 as a non-profit, nonpartisan organization, its purpose is to advance the public-pol-icy interests of its members and their consumers and to "provide member services to ensure adequate, reliable electricity at a reasonable price with the proper protection of the environment." APPA works closely with its members to:

- Assess their specific needs and design programs tailored to the unique interests of their organizations
- Provide in-house training
- Offer continuing-education units (CEUs) and professional-develop-ment hours (PDHs)

STATE UTILITY ASSOCIATIONS

State utility associations exist is almost every state. They are too numerous to list; however, the following are some outstanding examples.

National Rural Water Association

The National Rural Water Association (NRWA) is a nonprofit federation of state rural water associations. Its mission is to provide support services to its state associations, which serve more than 24,000 water- and wastewater-system members. Member state associations are supported by their water-and wastewater utility membership and offer a variety of state-specific programs, services, and member benefits.

Additionally, each state association provides training programs and onsite assistance in areas of operation, maintenance, finance, and governance. This organization helps state associations develop new rate schedules, set up proper testing methods, maintain or upgrade their operator licenses, and understand the ever-changing and complex governmental regulations. State associations have historically trained over 40,000 water- and wastewater-system personnel a year and provided over 60,000 onsite technical- assistance visits annually. More than 2600 ground-water protection plans have been adopted by local communities, and another 2300 are in the process of being adopted.

The Institute of Public Utilities

The Institute of Public Utilities supports the "informed, effective, and efficient regulation of utility network industries—electricity, natural gas, water, and telecommunications." The Institute fulfills its mission by providing to the regulatory-policy community integrative, interdisciplinary, and balanced educational programs and applied research on the institutions, theory, and practice of modern economic regulation. The Institute's analytical approach is designed to be "objective, broad-based, and informed by traditional and applied academic disciplines including law, business, engineering, communications, and the social sciences."

The Institute is an independent, self-supported, and nonprofit educational and research unit of Michigan State University, based in East Lansing, Michigan. Its core-program faculty members include nationally recognized university educators and expert practitioners across all utility sectors, issues, and perspectives. Established in 1965, the Institute has a tradition of service and responsiveness to the changing and challenging needs of regulatory professionals at the federal, state, and local levels as well as the international community. Institute programs are recognized by the National Association of Regulatory Utility Commissioners (NARUC) and provide continuing-education credits for regulatory professionals.

PROFESSIONAL SECURITY ASSOCIATIONS

ASIS International (ASIS) is the leading organization for security professionals. It has more than 33,000 members worldwide. Founded in 1955, ASIS is "dedicated to increasing the effectiveness and productivity of security professionals by developing educational programs and materials that address broad security interests, such as the ASIS annual seminar and exhibits, as well as specific security topics." ASIS also advocates the role and value of the security-management profession to business, the media, governmental entities, and the public. ASIS provides members and the security community with access to a full range of programs and services, and it publishes a national magazine, "Security Management."

PROFESSIONAL LAW ENFORCEMENT ASSOCIATIONS

The International Association of Chiefs of Police (IACP) was founded in 1893. The goals of this major association are to:

- Advance the science and art of police services
- Develop and disseminate improved administrative, technical, and operational practices and promote their use in police work
- Foster police cooperation and the exchange of information and experience among police administrators throughout the world
- Bring about recruitment and training in the police profession of qualified persons
- Encourage adherence of all police officers to high professional standards of performance and conduct

Throughout the past 100-plus years, the IACP has launched acclaimed programs, conducted important research, and provided exemplary programs and services to its membership around the world. Professionally recognized programs, such as the FBI Identification

Division and the uniform crime records (UCR) system, can trace their origins to the IACP. From spearheading national use of fingerprint identification to partnering in a consortium on community policing to gathering top experts in criminal justice, the government, and education for summits on violence, homicide, and youth violence, the IACP has had a major positive effect on the goals of law enforcement.

Chapter 12

FUTURE DIRECTIONS IN THE UTILITY/ENERGY INDUSTRY

Throughout this book, I have talked about essential improvements that are needed to secure our nation's utility/energy infrastructures. From time to time, I've also said that payment for these necessary enhancements and new systems can and should come from a variety of sources, including financial incentives from the federal government, increased fees for ongoing infrastructure services, and lower taxes for innovative infrastructure companies.

Here's an entirely different strategy that can supplement or perhaps even supplant incentives and fee adjustments. I recommend that the utility/energy industry take an entrepreneurial approach to meeting our nation's need for greater security. Specifically, companies providing essential utility/energy infrastructures can extend their services into the security field for the mutual benefit of their customers and themselves.

UTILITY-PROVIDED SECURITY SERVICES

Utility/energy infrastructure companies administer our nation's most important physical assets. As you know, without our national electric grid the country would grind to a halt. If that grid were down for weeks or longer, we as a country would be in grave physical, financial, and psychological peril. And the electric grid is only one of the nation's essential utility/energy infrastructures that must be secured.

To help fund the all-important process of securing these vital infrastructures, we need to gain a fresh perspective. These infrastructures are in a unique position to satisfy the security requirements of their industrial/commercial, educational, and homeowner customers. No other type of organization is better suited for this purpose. And utility/energy companies can ramp up efficiently by expanding the services they already offer.

Over the years, utility/energy infrastructure organizations have in fact become providers of community-based services, especially since deregulation. Because of their infrastructures and 24-hour operations, offering other services is a natural development. Following are some examples of community-based services offered across the United States:

- "One-call" underground-location services to telecommunication companies, cable companies, and others. This strength comes from leveraging in-house geographic information systems (GIS).
- Installation and monitoring of security services
- Medical-alert services for the elderly, especially in rural areas
- Twenty-four-hour dispatch services for smaller utilities
- Internet service providers (ISPs)
- Cable-television and satellite service providers
- In some areas of the United States, utility line crews are trained to alert law enforcement agencies to suspicious activities while on call or line patrol.

Utility/energy companies have excellent records of assisting local government agencies, state agencies, chambers of commerce, and other community-based organizations in relocating new business and expanding existing business. Utility/energy companies are natural providers of homeland security because they not only have infrastructures but also have the business experience to bring security to commercial and consumer customers. Utility/energy companies can make it happen.

The main objective in using the utility/energy industry as a provider of homeland security is first to secure the industry nationally and then to offer or expand their security programs to their key accounts and customers. As the utility industry increases the security of its facilities and infrastructures, it can help pay for these improvements by turning these cost centers into profit centers. Utilities can do this by extending their newly acquired expertise into services they offer their commercial/industrial accounts, college and school, airport, and homeowner customers. The following list offers basic programs the industry needs to establish in order to deter incidents from occurring or to alert authorities about potential security breaches:

- Physical security: Many utility/energy facilities are remote from large population areas and spread out. Because of those facts, some people claim that power plants, substations, power lines, mining facilities, and similar operations are impossible to secure. That is not true. By using modern technologies, most facilities can be secured and monitored on a 24/7 basis utilizing utility infrastructures that are already in place. Some examples include real-time remote sensing, remote surveillance, rapid response and deployment teams, biochemical monitoring of water supplies and facilities, and advanced physical-security tracking technologies that constantly monitor employees, vehicles, and company assets to include electronic barriers. Rapid response plans can be implemented by private guard services.

- Cyber security: Two main areas in every utility need protection, the computer systems in their headquarters and the SCADA systems, which control the power flow over transmission lines. These systems can be hacked by people with basic levels of computer skills who can be located in different parts of the world. If these systems are successfully penetrated by unauthorized individuals, the results can include brownouts, blackouts, or power surges that can severely damage client systems. Because of their unique configuration, SCADA systems are especially vulnerable. They require protections that must be enforced, awareness of potential problems, and users who are extremely cautious.

- Biochemical/anthrax security: In Chapter 2, I noted that biochemical agents could be introduced into power facilities by using utility infrastructures as weapons of mass destruction. Technologies are currently available that will sense the presence of these agents and enable utilities to shut off, divert, trap, or contain exposed contaminations before they become destructive.

- Security training: All management and staff at utility/energy facilities need to be trained in creating and sustaining a security culture to foster an environment of security. All eyes and ears need to constantly look for potential security breaches.

- Training will also prepare staff to respond to any actual or potential incidents in a proper and safe fashion and to be able to handle an emergency efficiently and effectively. There must also be training in new tools, resources, and techniques as they become available. In addition, the workforce needs to be educated so that all employees are more sensitized to potential problems and can prevent possible terrorist acts, fires, water damage, and loss from natural disasters, including hurricanes, tornadoes, and ice storms. Management needs to troubleshoot what would cause a broad or narrow shutdown. Some of these answers will result in industry-specific alerts and can include sensing, shunting, diverting, and turning off.

- Security background checks: It is critical to know whom we are employing at utility/energy facilities and the background of contractors who provide services to the industry. Mandatory background checking, including criminal background reviews, should be performed on all new hires. These background checks should be updated on an established schedule, perhaps yearly. When individuals are promoted, especially if they will occupy sensitive positions, background checks should be conducted. I realize that these executives might have been with the utility for years, but it is safest if "standard procedure" requires this special research upon promotion.

- Due diligence for contractors: The same due-diligence checks need to be done on contractors to determine, among other variables, if they have any affiliations with terrorist organizations. This essential analysis will prevent hiring or contracting with potential adversaries or businesses with conflicts of interest and avoid the possibility of the utility exposing itself and its customers to someone who might have inappropriate affiliations.

- Security notifications systems: These systems need to be in place to electronically alert federal, state, and local law enforcement, emergency services, and government agencies, as well as industry groups, immediately after a security incident has occurred. To help prevent incidents, we also need a better exchange of intelligence among industry groups and law-enforcement agencies.

- Crisis management and business-continuity planning: This will most probably be accomplished by outsourcing to experienced companies with excellent records of providing these services. You'll find a discussion of these subjects in Chapter 5.

Once these basic programs are in place at utilities, homeland-security offerings, branded in the name of the utility, can be provided to industrial and commercial accounts, federal and city government agencies, airport facilities, educational institutions at every level, and residential customers in all the areas the utility serves. Additional homeland-security services from the utility can include:

- Daily briefings via the utility website on national and regional security issues and threats
- Community emergency-preparedness plans
- Availability of supplies for subscribers to keep on hand in case of emergency and stockpiles of supplies in case of a long-standing emergency
- Latest information on types of biological/chemical threats and recommended preventive measures
- Real-time sources of information on which emergency services and criminal-justice community plans are available and where to get them
- Accurate and current lists of emergency-response and command and control personnel with correct contact information
- Recommended disaster-recovery/business-continuity programs
- News about technology advancements
- The utility can provide carefully selected security products and services and news and information about these products and services. These can include perimeter security, internal security, employee screening, computer security, risk management, and crisis management.

In these ways, utility/energy companies can become a very important part of our nation's homeland-security service. It is important to note, however, that the utility/energy industry would become a service that supports local and regional law-enforcement agencies and emergency services, as well as homeland-security operations. Each community would still rely on the criminal-justice system and local and national media as the primary sources for news and information.

It just makes good sense to use the infrastructures our utility/energy companies have in place. Keep in mind that that competitive fees will be charged by utility providers as they operate these services in ways that are similar to other parts of their business.

The profit from these efforts would be used to offset the costs of installing the systems that would help to secure our nation's utility/energy infrastructures. If done correctly, a very positive return on investment (ROI) could be realized in a short period of time.

Some utility/energy companies might want to supply services themselves. Others might want to subcontract through trusted national and regional technology companies and service providers in their local areas.

These subcontract relationships will not only make our communities safer; they can also add to employment, boosting local economies. And there are also technology companies that provide homeland-security programs specific to the utility/energy industry. Their technologies run on secure networks using real-time videoconferencing and monitoring all aspects of physical security, cyber security, and biochemical security. They also provide immediate notification alerts to all levels of government and law-enforcement agencies. These solutions are customized so a utility can plug and play with minimal installs, and they are especially designed to work in rural America.

For information about companies that can provide valuable homeland-security products and services, please go to www.nessgroup.com.

FEDERAL GOVERNMENT

Since 9/11, the federal government has been scrambling to get security programs in place to help protect U.S. citizens from other terrorist attacks. Although most government efforts have been focused on airport security, port security, and nuclear plants and on streamlining federal agencies, little attention has focused on the utility/energy industry. What's worse, the few efforts to date have been largely reactive in nature.

The real focus needs to be on preventing attacks by disgruntled employees and acts of terrorism, especially as they relate to the utility/energy industry. Being prepared to respond to an incident is critical. However, the tougher we make it for terrorists and other adversaries to attack, the less likely it is that incidents will occur.

The federal government is spending huge amounts of money on new technologies and security programs. That is excellent. Yet the answer to a number of our nation's security issues is right under the nose of the government decision makers. Our nation's utility/energy companies have infrastructures in place in every city, town, and rural area in America. In some cases, with only minor upgrades, these infrastructures can be used to help protect the citizens of the United States and our federal government from terrorist acts.

If you were to look at any city or town in America, local utility/energy providers serve almost everyone. They deliver, for example, electricity, gas, water, and telecommunications. To install security systems, biochemical monitoring capabilities, and early-warning systems, the existing infrastructures could be made cost-effective and reliable. In the future, if we are able to link all cities and towns together into the Department of Homeland Security (DHS) in Washington, DC, we could have instant national coverage and early warning of potential security issues.

If there were incentives for our utility/energy companies to become providers of homeland security, I believe most companies would participate, especially if there were liability waivers put in place by the federal government. These waivers would also provide the opportunity for the DHS in Washington to start a much-needed national movement to protect our nation's most valuable assets, our utility/energy infrastructures.

To help secure these utility/energy infrastructures, the federal government must become proactive in the following ways:

- Develop national security standards for the utility/energy industry. To accomplish this, the federal government needs to involve the Department of Energy, the Federal Energy Regulatory Commission, state regulators, energy providers, and security experts in the utility/energy arena.
- Supply grants for vulnerability assessments and funding to help fix the problems
- The Department of Energy needs to complete crafting the standards that the utility/energy industry sorely needs.
- Congress must finalize an energy bill that includes tax incentives to encourage and promote those utility/energy companies that secure their facilities. It should include minimum standards of compliance.
- Congress must provide all necessary liability waivers to utility/ energy companies that put security programs in place and market them to

their customer base. The liability waivers from the DHS are part of the Support Anti-Terrorism by Fostering Effective Technologies (SAFETY) Act. This Act is a good start because it offers special liability and immunity protections to companies that develop innovative technologies that are designed to help in the war against terrorism.

- Congress must also offer utility/energy companies safeguards that provide specific protections from the Freedom of Information Act.

- The Department of Energy needs to organize the national grid systems so they operate in a stand-alone manner in many regions of the United States. We need failsafe grids with re-shunting capabilities. This strategy is safer in terms of security than having large, regional grids that, if taken down, disrupt major sections of the nation and have the potential for cascading disruptions. Texas, for example, has its own grid. If problems developed in the Texas statewide system, they would not affect transmissions in neighboring states.

- Congress needs to provide incentives to our best technology companies so they are motivated to develop new technologies that protect our utility/energy infrastructures.

- The DHS needs to supply energy providers with real-time intelligence as it relates to their industry. Our nation's intelligence agencies collect a lot of information from many sources, including the utility/energy industry. Once the information is collected, it gets classified in Washington, DC, so the federal government cannot share it with utilities. We need to develop a program where at least two individuals at every utility have a security clearance so they can receive intelligence materials as they relate to the utility industry. Only in that way can all companies in the utility/energy industry plan and take proper security measures to avoid possible attacks.

- Congress must support an energy tax to help pay for enhancements, as well as better notification plans and methods. We also need to encourage state regulators to allow rate increases for security.

Commercial and Industrial Accounts

As our nation's utility/energy infrastructures are becoming more secure, utility/energy companies can offer homeland-security programs to the commercial and industrial accounts they serve. This can be established so that energy companies earn new revenues that help pay for increased security for their facilities and infrastructures, profit from their security offerings, and, in return, provide protection to all major industrial and commercial accounts as well as homeowners in their service areas.

Those served would include military and government facilities, hospitals, airports, banks and other financial institutions, telecommunications companies, technology firms, and many, many more. Ask most people if they trust their utility/energy company to provide quality services. Almost always you will get a positive response. Here are some ways for utility/energy companies to extend their services while benefiting the communities they serve:

- Utility/energy companies can provide communications plans for each employee at a subscriber company to act upon in case of terrorism.
- At the local and regional level, utility/energy companies can offer in-person and written educational programs that explain the different actions to take dependent upon different types of attacks—chemical, biological, fire, and nuclear. Initial information about these programs can be distributed in monthly mailings.
- They can subcontract with excellent providers of services and recommend those providers to customers. Examples can include alarm companies, guard services, personnel screening services, computer services, as well as risk- and crisis-management firms.
- If they don't want to subcontract, utility/energy companies can provide their customers with lists of security product and service vendors who have been approved by the General Services Administration (GSA).

Colleges and Schools

Assume homeland-security programs are in place or growing at a utility and a major effort is underway to provide commercial and industrial accounts with such services. Next comes the opportunity to aid schools and colleges. The tragic terrorist attack in 2004 at a school is Beslan, Russia makes this extension of security a priority. Intelligence has discovered that schools have also been monitored by terrorists in specific parts of the United States.

Once a national-homeland security effort is established through our utility/energy companies, the federal government will need to offer incentives to provide no-charge or highly subsidized programs to our schools and colleges. Most schools have very little security, and most of what is offered focuses on physical security, which is usually only monitored within the schools themselves. Security needs to be expanded and monitored outside the school as well as internally. This will help to ensure detection and immediate response should an incident occur.

Local utility companies can meet many of the homeland-security services of our schools. Here are some of the many ways that utilities can help secure our educational institutions:

- Utility/energy companies can provide security plans for actions to take in case of a terrorist attack. They can send these plans to administrators at colleges and school districts, who can distribute them to appropriate faculty and staff.

- Utility/energy companies or their subcontractors can offer in-person and written educational programs that present the actions to take for different types of incidents, including terrorist attacks on electrical, chemical, biological, and nuclear facilities, as well as fires and natural disasters.

- Utility/energy companies can provide colleges and school districts with lists of licensed providers of security services. Examples can include alarm companies, guard services, personnel-screening services, computer services, as well as risk- and crisis-management firms.

- Utility/energy companies can provide monitoring services. They can monitor alarms and meters and, when necessary, report incidents to educational officials and to law-enforcement personnel.

HOMEOWNERS

The homeland-security offering for utility/energy companies, industrial and commercial accounts, and universities, colleges, and schools, should eventually be extended to homeowners.

Utilities have been and continue to be incentivized in many ways for undertaking the responsibility of improving power quality, cutting power demand, and reducing pollution with green pricing programs. These incentives take many forms: tax credits, favorable regulatory treatment on returns, and profit sharing with customers, just to name a few. These tried and proven incentives should be deployed to help utilities extend the reach of homeland-security services to their many markets.

I want to note that for years a significant number of utilities have been offering successful power-quality services and solutions primarily to increase customer satisfaction. A recent study by Atlanta-based Chartwell, Inc., found that of the 72 electric utility companies providing products and services to commercial and industrial customers, two-thirds of the utilities offered power-quality services at no charge. These services included diagnostics and suggestions for mitigation. "Profit is not the underlying motivation. Increasing customer satisfaction is," states the Chartwell report. With these and many other long-established customer connections, it would be relatively inexpensive and very efficient to add a homeland-security component.

Think about it: Utilities really are the right entities to offer homeland-security services. They are ideal conduits for making our nation a safer place to live., since they:

- Can't pick up and move out of state
- Are tightly regulated in terms of what can be charged for services
- Know their customers better than many other types of businesses in a geographically bounded area
- Already have many other service pathways into local and regional accounts

Is there a good business case for utilities to offer homeland-security services? Absolutely. The cost for a utility to enter into this business is minimal, considering all the previously funded expenses in such areas as market research, sales staffs, customer-education programs, financing programs (leasing, debt financing, etc.), and bundled offerings. These core competencies have been honed for many years with prior experience in power quality, energy conservation, and green pricing initiatives. The built-in efficiencies translate to a "win-win" proposition, especially as it relates to providing security technology and monitoring products and services. Customers could purchase homeland-security services at economical prices and utilities can earn a fair rate of return for their investors. In addition, there are economic-development aspects of professional job creation and major benefits that will result from the increased levels of security, which will attract new business to the area.

Here are some of the security services utilities can offer homeowners:

- Provide information on all perimeter and internal alarm-system companies in the area
- Informational programs regarding how to respond if different types of security events occur and how to talk with children so they will be confident while also prepared

- Lists of supplies to have on hand to be prepared for a variety of events. A list of the best local stores to buy these supplies would be good for the community and a goodwill gesture on the part of utilities.

Our nation's utility/energy infrastructures can and must be made much more secure. To achieve this urgent requirement, government at federal and state levels needs to be proactive. But more importantly, so do members of management of our utility/energy companies. Our country's most valuable physical assets are in their trust. It is in our nation's vital interests for us all to work together to protect these critical infrastructures from terrorist attack and make America safer.

Chapter 1   Appendix

United States General Accounting Office

GAO

Report to the Chairman, Senate
Committee on Governmental Affairs

November 2003

ELECTRICITY RESTRUCTURING

2003 Blackout Identifies Crisis and Opportunity for the Electricity Sector

G A O
Accountability * Integrity * Reliability

GAO-04-204

Contents

United States General Accounting Office
Washington, DC 20548

November 18, 2003

The Honorable Susan M. Collins
Chairman, Committee on Governmental Affairs
United States Senate

Dear Chairman Collins:

The August 14, 2003, electricity blackout—the largest in the nation's history—affected millions of people across eight northeastern and midwestern states as well as areas in Canada. In some areas, power was restored in hours, while in others power was lost for several days. The blackout intensified concerns about the overall status and security of the electricity industry at a time when the industry is undergoing major changes and Americans have a heightened awareness of threats to security.

Because of these widespread concerns and the broad institutional interest of the Congress, we (1) highlighted information about the known causes and effects of the blackout, (2) summarized themes from prior GAO reports on electricity and security matters that provide a context for understanding the blackout, and (3) identified some of the potential options for resolving problems associated with these electricity and security matters.

Over the past several weeks, GAO staff briefed numerous congressional staff on its observations. In response to your request, we prepared this overview to accompany the slides used in these presentations. Appendix I presents the latest briefing slides in their entirety. Our briefing is based largely on reports we previously issued on a range of electricity issues along with updated information obtained from the Department of Energy (DOE), the North American Electric Reliability Council, and operators of the transmission system in the blackout region. The information presented is intended to place the electricity blackout in the broader context of long-term issues affecting the sector. The options presented do not encompass a complete set of all possible options but do represent ideas that merit consideration as the nation moves forward to address this important issue.

Summary

While the root cause of the blackout has not yet been conclusively established, a recent DOE report describes a sequence of events that culminated with the outage. A series of power plants and transmission lines went offline beginning at about noon eastern daylight time because of instability in the transmission system in three states. The loss of these plants and transmission lines led to greater instability in the regional power transmission system, which—4 hours later—resulted in a rapid cascade of additional plant and transmission line outages and widespread power outages. The blackout affected as many as 50 million customers in the United States and Canada, as well as a wide range of vital services and commerce. Air and ground transportation systems shut down, trapping people far from home; drinking water systems and sewage processing plants stopped operating; manufacturing was disrupted; and some emergency communications systems stopped functioning. The lost productivity and revenue have been estimated in the billions of dollars. A joint U. S.-Canadian taskforce is seeking to identify the root cause of the failures and plans to issue an interim report in November 2003.

Over the past several years, our work on the electricity sector has resulted in numerous findings, conclusions, observations, and recommendations. Based on this prior work, we highlight three themes on electricity and security matters in our briefing and lay out some of the potential options to consider in addressing problems in these areas.

Specifically:

- Electricity markets are developing, but significant challenges remain. Our work has shown that while the electricity sector is in transition to competitive markets, the full benefits of these markets will take time and effort to achieve. For example, we found that the separate development of wholesale and retail electricity markets, which is occurring as part of the electricity industry shifts from regulated to competitive markets, limits the industry's ability to achieve the benefits of competition. The separate development of these markets reduces or eliminates retail consumers' incentive or ability to respond to market signals that supplies are tight. Consumers do not respond because the retail prices they see are set by state regulators and do not reflect actual market conditions. This lack of consumer response becomes particularly important during periods of high demand for electricity, such as hot summer afternoons, when total electricity use approaches the total amount of available generation. Efforts to promote various types of demand response, such as those that link customers' electricity consumption with prices, may offer one option for improving this situation. We are exploring this issue in more depth in response to your request. Other issues raised by our work in this area are presented in slides 14 through 18 of the briefing.

- Oversight of markets and reliability needs more attention. The ongoing transition to competitive markets, or "restructuring" of electricity markets, has dramatically changed how the Federal Energy Regulatory Commission (FERC) needs to oversee these markets and the information it needs to do so. In order to monitor current market conditions to ensure fair competition, for example, FERC needs to access market information on wholesale transactions and the operation of electric generating plants, among other things. Our work shows that FERC's oversight efforts are improving, but it continues to be hampered by a number of challenges. In particular, we noted that FERC had previously not clearly defined its role in monitoring the market, faced gaps in information due to limitations in its jurisdictional authority, relied on third-party data to perform regulatory functions, and had limited enforcement authority. In addition, we pointed out that FERC faced human capital challenges to acquire and develop the staff knowledge and skill it needs to effectively regulate and oversee today's electricity market. Because restructuring has changed the types of information regulators need, we have previously recommended that FERC demonstrate what additional information it needs, describe the limitations it faces without such information, and ask the Congress for authority to collect it. One option for congressional action in this area includes providing FERC with authority to gain access to needed data relating to reliability and markets. Other issues raised by our work in this area are presented in slides 19 through 26 of the briefing.

- Security for critical infrastructure is of growing importance. Our work has shown that a reassessment of the security of the nation's physical infrastructure as well as that of related information technology and control systems should be undertaken. Often, security measures have been added after the infrastructure is in place, which is costly and creates potential conflicts between security and efficiency. Therefore, it may be better to integrate sufficient security measures for these critical systems, particularly in a post-September 11th environment, into the planning for new construction or the upgrading of existing infrastructure, rather than viewing them as later add-ons. Our work has also raised concerns about the increasing reliance on information technology and control systems, which are potentially vulnerable to cyber attack, including the systems used in the electricity sector. As part of our work, we have found that cyber attacks against these systems could be used to cause damage or complicate the response to a physical attack. One option to help address this problem would be to increase the focus on research and development and other related activities, including the use of currently available technologies and vulnerability assessments, aimed at enhancing national capabilities to respond to cyber-security issues. Other aspects of our work in this area are presented in slides 27 through 29 of the briefing.

Whatever the ultimate cause of the blackout, our work has shown that a number of significant challenges remain for the electricity sector. We recognize that many issues surrounding the restructuring of the electricity industry are complicated and that solutions involve complex policy tradeoffs for the Congress that will undoubtedly take time to fully resolve. GAO stands ready to provide any analytical assistance the Congress may need in this important long-term endeavor.

We conducted our work in accordance with generally accepted government auditing standards.

We are providing copies of this report to other appropriate congressional committees as well as DOE and FERC. The report will be available at no charge on the GAO Web site at http://www.gao.gov.

If you or your staff have any questions about this report, please contact me at (202) 512-3841. Major contributors to this report included Mary Acosta, Dennis Carroll, Dan Haas, Randy Jones, Mike Kaufman, Jon Ludwigson, and Barbara Timmerman.

Sincerely yours,

Jim Wells
Director, National Resources and Environment

Appendix I: 2003 Blackout Identifies Crisis and Opportunity for the Electricity Sector

2003 Blackout

Crisis and Opportunity for the Electricity Sector

Contents

Information on the Blackout

Background on Electricity and Electricity Restructuring

Themes from Prior GAO Work

Potential Options

What We Considered in Developing the Briefing

Examined information from NERC, FERC, DOE, several ISOs

Examined prior relevant GAO work (reports, testimony, and briefings) covering a range of areas

- Development of energy markets
- Regulatory oversight of energy sector
- Homeland security and physical security of critical infrastructure
- Information technology and cyber security

Information on the Blackout
Events Preceding the Blackout

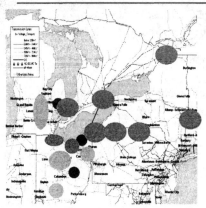

Root cause not known, evaluation ongoing

Current analysis points to a series of line and plant outages (times in Eastern Daylight Time)

- Midday, 3 large power plants go down (Michigan, mid-Ohio, Cleveland)
- 2-3 PM 4 large (345 KV) transmission lines tripped
- Power swings noted in Canada and Eastern United States
- 4:10 many power plants and large transmission lines trip, mostly in Michigan and Ohio
- 4:11 cascade of trips sever New York east and west, PJM and New England separate

Grid operator (Midwest Independent System Operator) experienced significant communication problems during event

Note: Color coded dots depict approximate power line and power plant outage locations.
Source: GAO analysis of data provided by the Department of Energy and Platt's PowerMAP.

Information on the Blackout
Largest Blackout in U.S. History

Note: Locations and boundaries are approximate.
Source: GAO analysis of information provided by the Department of Energy.

Significant electricity outage
- 8 states and 2 Canadian provinces
- 61,800 MW of demand
- 50 million customers (estimated)
- About 102 plants (22 nuclear plants)

Wide-ranging impacts
- Water systems shut down
- Raw sewage dumped
- Air and ground transportation halted
- Gas stations and refineries closed
- Cellular networks interrupted
- 911 communications interrupted
- Manufacturers shut down

Investigations underway
- US-Canadian team
- U.S. House—Energy and Commerce
- U.S. House—Homeland Security
- State government investigations

Background
Electricity Sector

Vast infrastructure
- 3 major U.S. systems (East, West, and Texas)
- Vast network (~2,100 large power plants, ~150,000 miles of electricity lines)
- 4 key functions (generation, transmission, distribution, system operations)

Integrated and coordinated system
- Electricity must be produced and consumed at almost the same time
- A change in one area of the grid can affect other areas, almost instantaneously
- Infrastructure (such as power plants and transmission grid) has limited capacity
- Local and regional supply and demand must be balanced to avoid blackouts
- Overall demand increases about 2-3% per year, but can vary widely by region

Regulated sector
- Federally at FERC (wholesale sales, interstate transmission for "public utilities", generally excludes Alaska, Hawaii, and most of Texas)
- States through state commissions (retail sales, intrastate transmission)
- Some entities largely unregulated (cooperative, municipal, and other owner-serving entities)

Background
Local Networks Became 3 Synchronized U.S. Systems

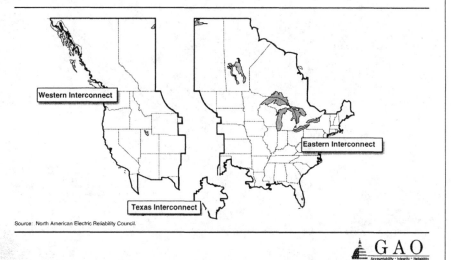

Western Interconnect

Eastern Interconnect

Texas Interconnect

Source: North American Electric Reliability Council.

GAO
Accountability · Integrity · Reliability

Background
Vast Network of Electricity Lines and Plants

Source: GAO Analysis of Data Provided by Platt's.

GAO

Background
Electricity Sector Has 4 Distinct Functions

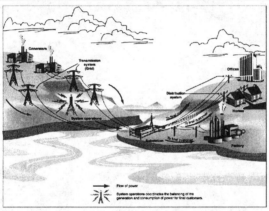

Source: GAO.

Background
System Must be Balanced Throughout Day

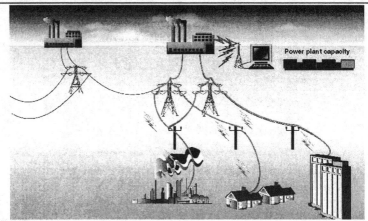

Source: GAO.

Background
Restructuring Goal: Competition Leading to Benefits

Increase competition
- Increase numbers of buyers and sellers
- Provide information to enable consumers to make choices
- Enable sellers to enter and exit market in response to information

Provide benefits to consumers
- Lower prices
- Access to new services
- Increased efficiency
- More innovation

Maintain or enhance reliability

Background
Electricity Restructuring Is Changing the Sector

Businesses shifted from emphasizing regulation to emphasizing markets
- Opened generation to competition, 3 other functions remain regulated
- Companies now bear risks previously borne by ratepayers

Physically, opened access to transmission lines to utilities and new entrants
- As a result, multiple entities now interact to perform the key functions needed to deliver electricity to consumers

Regulators shift away from planning system and setting rates
- FERC is shifting from setting rates (reactive) to designing and monitoring markets in real-time (proactive)
- States are moving away from central role in planning and rate setting to leaving private entities to determine what to build and what to charge

Themes from Prior GAO Work
GAO Work Focused Around Three Key Areas

Electricity markets developing, but challenges remain
Regulatory oversight needs more attention
Security of critical infrastructure increasingly important

Themes from Prior GAO Work
Electricity Markets Developing, but Challenges Remain

Adequate infrastructure essential for reliability and competitive prices

Grid becoming more regional, but key authorities remain with states and localities

Little understanding of options during market events or emergencies

Uncertainties limit investment in new infrastructure

Themes from Prior GAO Work
Electricity Markets Developing, but Challenges Remain

Adequate infrastructure essential for reliability and competitive prices

- Growing electricity demand places pressure on grid and supply sources
- Restructuring increases use of the grid
- Numerous parts of U.S. witnessing increasing congestion of transmission system

Note: Arrows in this illustration show the location and direction of current transmission congestion.

Source: North American Electric Reliability Council.

Themes from Prior GAO Work
Electricity Markets Developing, but Challenges Remain

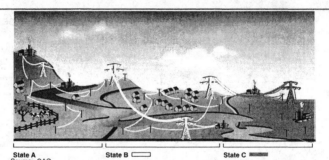

State A
Source: GAO.   State B ⬚   State C ▬

Grid becoming more regional, but authorities remain with states and localities
- Infrastructure serves regional needs
- Increasingly, problems in one state affect neighboring states
- Authority to add new infrastructure remains with states

Themes from Prior GAO Work

Electricity Markets Developing, but Challenges Remain

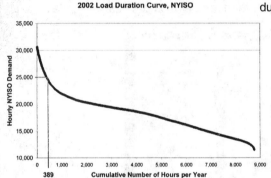

2002 Load Duration Curve, NYISO

Source: GAO analysis of New York Independent System Operator data.

Little understanding of options during times of shortage or crisis

- Highest levels of demand only reached during a small percentage of the hours of the year
  - For example, the New York ISO exceeds 25,000 MW during <5% of hours
- Consumer response to scarcity, such as when wholesale prices rise, is missing due to lack of information and incentives
- Backup generation not inventoried, limits on use
- Use of emergency power generally involves environmental and other challenges

Themes from Prior GAO Work
Electricity Markets Developing, but Challenges Remain

Attracting private investment requires balancing risk and profits

•Market
•Business
•Environmental regulation
•9/11, terrorism
•Enron, investment
•Federal regulation
•State regulation

•Market prices v. costs
•Higher cost of capital
•Shorter recovery periods

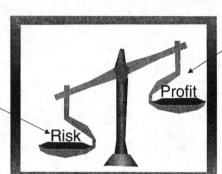

Source: GAO.

Themes from Prior GAO Work
Regulatory Oversight Needs More Attention

Monitoring of reliability inadequate
Limited jurisdiction and varied rules hinder market development
FERC faces challenges in market oversight

Themes from Prior GAO Work
Regulatory Oversight Needs More Attention

Monitoring of reliability inadequate

	FERC	States	NERC
Monitoring Authority	Transmission Wholesale Markets	Distribution Retail Markets	None (Voluntary Membership)
Reliability Monitoring	None	Varies (23 of 40 states*)	Nationally (Voluntary Compliance)

*National Regulatory Research Institute

Source: GAO.

Themes from Prior GAO Work
Regulatory Oversight Needs More Attention

Limited jurisdiction and varied rules hinder market development

- FERC lacks jurisdiction over some entities resulting in patchwork of restructured and regulated wholesale markets
- Varied wholesale market rules may limit development of a competitive market
- Limited implementation of retail restructuring may limit development of wholesale markets
- FERC protects consumers through its oversight of wholesale markets and through its oversight of fair access to transmission lines

Appendix I: 2003 Blackout Identifies Crisis
and Opportunity for the Electricity Sector

Divided Jurisdiction
FERC Lacks Jurisdiction Over Some Entities

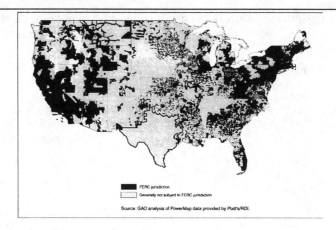

FERC jurisdiction
Generally not subject to FERC jurisdiction

Source: GAO analysis of PowerMap data provided by Platt's/RDI.

Divided Jurisdiction
FERC Lacks Jurisdiction Over Some Entities (cont.)

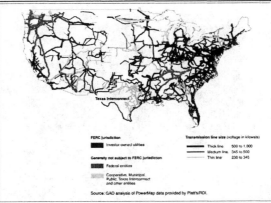

FERC jurisdiction

■ Investor-owned utilities

Generally not subject to FERC jurisdiction

■ Federal entities

▒ Cooperative, Municipal, Public, Texas Interconnect and other entities

Transmission line size (voltage in kilowatts)

——— Thick line   500 to 1,000
——— Medium line   345 to 500
——— Thin line   230 to 345

Source: GAO analysis of PowerMap data provided by Platt's/RDI.

Notes: Data for transmission lines reflect primary ownership--some lines have multiple owners.

Federal entities include the Bonneville Power Administration, the Tennessee Valley Authority, the Western Area Power Administration, and others.*

GAO
Accountability * Integrity * Reliability

Divided Jurisdiction
Varied Wholesale Market Rules

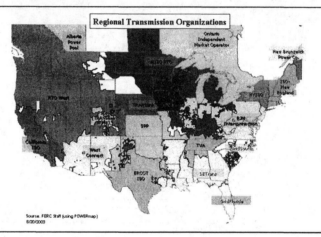

Divided Jurisdiction
Limited Implementation of Retail Restructuring

- Restructuring active
- Restructuring delayed
- Restructuring suspended
- Restructuring not active

Source: GAO Analysis of Energy Information Administration data.

GAO
Accountability * Integrity * Reliability

Themes from Prior GAO Work
Regulatory Oversight Needs More Attention

FERC faces market oversight challenges

Effective Oversight

Enforcement

3rd Party Data

Information Gaps

Human Capital

Role

Source: GAO.

Themes from Prior GAO Work
Security of Critical Infrastructure Increasingly Important

Embedding homeland security principles as integral part of investment in infrastructure and business processes is important

Increasing reliance on information technology requires attention to cyber-security

Themes from Prior GAO Work
Security of Critical Infrastructure Increasingly Important

Embedding homeland security principles as integral part of investment in
infrastructure and business processes is important

- Designing it in up-front [most cost-effective]
- Not bolting it on afterwards [potential conflicts between security and efficiency]

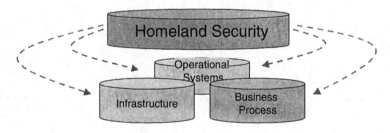

Source: GAO.

Themes from Prior GAO Work
Security of Critical Infrastructure Increasingly Important

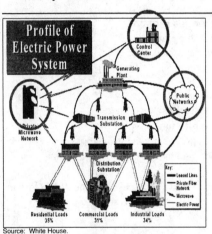

Source: White House.

Increasing reliance on information technology requires attention to cyber-security

- SCADA/control systems perform vital functions in electric power and other industries
- SCADA/control systems and networks are potentially vulnerable to attack, by terrorists or others
- Electricity sector making progress in implementing critical infrastructure protection (CIP) measures

Potential Options
Issues Needing Resolution and Options to Address Them

Energy markets developing, but challenges remain
- Minimal consumer responsiveness to market
- Lack of "backstop authority" for transmission siting
- Divided regulatory authority
- Lack of uniformity in market rules
- Uncertainty about future of restructuring
- Uncertainty over how to pay for transmission upgrades

Oversight and monitoring needs more attention
- Lack of enforceable reliability rules (NERC or federal reliability organization)
- Limited access to needed reliability and market data for regulators

Security of increasing importance
- Limited incentives for participation in federal critical infrastructure protection activities
- Homeland security concerns not fully considered
- Weaknesses in cybersecurity

Potential Options for Improving Consumer Responsiveness to Market

Issue—Minimal consumer responsiveness limits the development of markets and reduces reliability

- Infrastructure limited, resulting in scarcity during periods of high demand
- Consumers generally do not respond to this scarcity
- Lack of response increases costs, raises price volatility, and reduces reliability

What FERC has done

- In its standard market design, FERC has proposed to introduce one type of demand response

What GAO has done

- Identified that the lack of demand response limits the benefits of restructuring
- Examining the issue of demand response, release expected 2004

Options for congressional action

- Encourage the use of demand response tools in retail markets
- Promote the development of metering needed to implement demand response
- Educate utilities and consumers about the benefits of demand response

Potential Options for Providing "Backstop Authority"

Issue—Transmission serves regional needs and is essential to reliability, but varied approaches present challenges for upgrading system
- Reliability and markets depend on adequate infrastructure
- States and localities have regulatory authority over siting plants and lines
- Varied approaches to siting can result in contentious, uncertain, and time-consuming efforts to add infrastructure
- Uncertainty and delays may reduce investment in new infrastructure

What GAO has done
- Described FERC authorities for natural gas pipelines
- Described Colorado backstop statute for transmission lines

Options for congressional action
- Consider empowering regional entity to resolve disputes
- Consider FERC backstop authority (ability to grant eminent domain) if companies and states cannot reach agreement

Potential Options Addressing Divided Regulatory Authority

Issue—Divided regulatory authority limits development of market and reliability of the system

- Entire network interconnected—points on grid separated by milliseconds
- Lack of coordination between the development of wholesale and retail markets limits the potential benefits
- A patchwork of rules now governs electricity markets
  - Only about 75% of the wholesale market subject to FERC's authority
  - Varied state approaches to regulating/restructuring retail markets
- Lack of consistency limits trade and slows development of competition

What FERC has done

- FERC seeking consensus through outreach with states and regions

Options for congressional action

- Clarify FERC's authority over all wholesale markets and transmission lines
- Determine if FERC should oversee some aspects of retail markets
- Explore how to better link wholesale and retail markets

GAO
Accountability · Integrity · Reliability

Potential Options for Enhancing Uniformity of Market Rules

Issue—Lack of uniformity in market rules limits development of market and places reliability at risk

- FERC has approved different sets of market rules for different areas
- Presence of different rules limits trade, and the development of competition
- Presence of different rules makes investment in new infrastructure more risky

What FERC has done

- In 2002, FERC proposed standard market design that would make all FERC jurisdictional wholesale markets operate under a single set of rules

Option for congressional action

- Make market rules regional and move toward standardization—markets cannot solve this problem

Potential Options for Reducing Uncertainty About Restructuring

Issue—Lingering uncertainty over the future of restructuring is limiting needed investment and raises costs to consumers/investors
- Long-standing federal debate over future of restructuring creates uncertainty
- Actions on state restructuring of retail markets vary and the future is uncertain
- Uncertainty limits and/or delays investment and raises costs

What FERC has done
- Issued orders restructuring wholesale markets and proposals outlining plans

What GAO has recommended
- FERC should report on the performance of restructured electricity markets, collecting data needed to evaluate both wholesale and retail elements
- FERC should report annually to Congress to identify emerging issues and impediments to reaching goal of achieving competitive wholesale markets

Options for congressional action
- Reduce uncertainty about restructuring's future—identify milestones or timeline
- Increase incentives for investment

Potential Options for Determining How to Pay for Transmission Upgrades

Issue—Transmission upgrades vital for reliability and market development, but determining who should pay is difficult

- Transmission serves regional needs and promotes reliability
- Electricity often crosses jurisdictions and travels varied paths
- Determining who benefits and who should pay for additions is difficult
- New lines needed; infrastructure old and congestion increasing
- Private investment faces significant hurdles

What GAO has recommended

- FERC should standardize interconnection agreements to reduce uncertainty
- FERC should clarify how costs of transmission upgrades associated with interconnection should be allocated

Options for congressional action

- Increase regulated rate of return for investments in transmission
- Consider providing additional incentives, such as direct tax incentives or accelerated depreciation

Potential Options for Enhancing Reliability Rules

Issue—Voluntary nature of NERC membership and provision of data leaves gaps in information and threatens reliability

- Many new electricity producers not members in NERC and do not provide comprehensive operations data to it
- Nonmembers are not required to follow NERC guidelines for maintaining reliability

Options for congressional action

- Require mandatory compliance with NERC reliability guidelines
- Consider the creation of a new federal reliability organization

Potential Options for Improving Regulator Access to Needed Data

Issue—Restructuring changes what types of information regulators need

- FERC has shifted from approving cost-based rates to monitoring market performance
- FERC needs greater access to real-time market data to assess market performance and to identify abuses of market power
- FERC needs data on the operation of power plants to monitor reliability as well as assess performance of markets

What GAO has recommended

- FERC should demonstrate what additional information it needs
- FERC should describe the limitations it faces without having adequate information
- FERC should ask the Congress for authority to collect the information that it needs

Options for congressional action

- Give FERC authority to collect, or have access to, needed market data
- Give FERC access to reliability data

Previous GAO Recommendations Related to CIP Participation Incentives

Issue—Participation in federal CIP efforts critical

- Federal policy encouraged voluntary industry participation in critical infrastructure protection (CIP) activities, including operation of industry information sharing and analysis centers (ISAC's) and tasked lead agencies to assess the need for incentives for industry participation

What has been done

- Industry progress reported in fulfilling suggested ISAC activities
  - Establishing baseline statistics and patterns
  - Serving as clearinghouse within/among various sectors
  - Providing a library of historical data for private sector and government

What GAO has recommended

- DOE to assess the need for public policy tools (e.g., regulation, grants, tax incentives) to encourage increased industry CIP activities
- Identify additional actions needed to improve the quality and quantity of information being provided by the ISACs,

Potential Options for Addressing Homeland Security Concerns

Issue—Changing security threat

- Post-9/11
- Who's in charge and who pays?
- Measures to improve resiliency, redundancy, remote monitoring, fail-safe and restoration capabilities promote both greater efficiency as well as security

Options for congressional action

- Consider homeland security as integral and compatible to improving safety and reliability of systems when making investment decisions and designing improvements
- Consider distributed power sources, diversity in energy sources, improved conservation measures and remote sensing as ways to enhance both system safety as well as counter terrorist threats

Potential Option for Addressing Cybersecurity Weaknesses

Issue—Electricity sector faces increasing cyber threats

- Supervisory control and data acquisition (SCADA)/ control systems, critical to the electric industry, may not have adequate security
- Cyber attacks against these systems could be used to cause damage or complicate the response to a physical attack

What has been done

- A number of federal and private-sector efforts are underway to study cybersecurity issues
  - Including developing standards and performing research and development

Options for congressional action

- Increase focus on research and development and related efforts to address cybersecurity issues

Prior Relevant GAO Work

Natural Resources and Environment Team
- *Bonneville Power Administration: Long-Term Fiscal Challenges,* GAO-03-918R
- *Energy Markets: Additional Actions Would Help Ensure That FERC's Oversight and Enforcement Capability Is Comprehensive and Systematic.* GAO-03-845
- *Electricity Markets: FERC's Role in Protecting Consumers,* GAO-03-726R
- *Electricity Restructuring: Action Needed to Address Emerging Gaps in Federal Information Collection,* GAO-03-586
- *Lessons Learned From Electricity Restructuring:Transition To Competitive Markets Underway, but Full Benefits Will Take Time and Effort To Achieve,* GAO-03-271
- *Restructured Electricity Markets: California Market Design Enabled Exercise of Market Power,* GAO-02-828
- *Energy Markets: Concerted Actions Needed by FERC to Confront Challenges That Impede Effective Oversight,* GAO-02-656
- *Restructured Electricity Markets: Three States' Experiences in Adding Generating Capacity,* GAO-02-427
- *California Electricity Market Options for 2001: Military Generation and Private Backup Possibilities,* GAO-01-865R
- *Energy Markets: Results of Studies Assessing High Electricity Prices in California,* GAO-01-857

Information Technology Team
- *Critical Infrastructure Protection: Challenges for Selected Agencies and Industry Sectors,* GAO-03-233
- *Information Security: Progress Made, But Challenges Remain to Protect Federal Systems and the Nation's Critical Infrastructures.* GAO-03-564T
- *High Risk Series: Protecting Information Systems Supporting the Federal Government and the Nation's Critical Infrastructures,* GAO-03-121
- *Combating Terrorism: Selected Challenges and Related Recommendations,* GAO-01-822

$\triangleq$ G A O
Accountability • Integrity • Reliability

GAO's Mission

The General Accounting Office, the audit, evaluation and investigative arm of Congress, exists to support Congress in meeting its constitutional responsibilities and to help improve the performance and accountability of the federal government for the American people. GAO examines the use of public funds; evaluates federal programs and policies; and provides analyses, recommendations, and other assistance to help Congress make informed oversight, policy, and funding decisions. GAO's commitment to good government is reflected in its core values of accountability, integrity, and reliability.

Obtaining Copies of GAO Reports and Testimony

The fastest and easiest way to obtain copies of GAO documents at no cost is through the Internet. GAO's Web site (www.gao.gov) contains abstracts and full-text files of current reports and testimony and an expanding archive of older products. The Web site features a search engine to help you locate documents using key words and phrases. You can print these documents in their entirety, including charts and other graphics.

Each day, GAO issues a list of newly released reports, testimony, and correspondence. GAO posts this list, known as "Today's Reports," on its Web site daily. The list contains links to the full-text document files. To have GAO e-mail this list to you every afternoon, go to www.gao.gov and select "Subscribe to e-mail alerts" under the "Order GAO Products" heading.

Order by Mail or Phone

The first copy of each printed report is free. Additional copies are $2 each. A check or money order should be made out to the Superintendent of Documents. GAO also accepts VISA and Mastercard. Orders for 100 or more copies mailed to a single address are discounted 25 percent. Orders should be sent to:

U.S. General Accounting Office
441 G Street NW, Room LM
Washington, D.C. 20548

To order by Phone: Voice: (202) 512-6000
                    TDD: (202) 512-2537
                    Fax: (202) 512-6061

To Report Fraud, Waste, and Abuse in Federal Programs

Contact:

Web site: www.gao.gov/fraudnet/fraudnet.htm
E-mail: fraudnet@gao.gov
Automated answering system: (800) 424-5454 or (202) 512-7470

Public Affairs

Jeff Nelligan, Managing Director, NelliganJ@gao.gov (202) 512-4800
U.S. General Accounting Office, 441 G Street NW, Room 7149
Washington, D.C. 20548

PRINTED ON RECYCLED PAPER

Chapter 3   Appendix

- **LABORATORIES AND RESEARCH FACILITIES**
  Homeland Security Centers of Excellence

- **HOMELAND SECURITY URGES SMALL AND MEDIUM-SIZED BUSINESSES TO TAKE STEPS TO PREPARE FOR EMERGENCIES**

- **2004 COUNTERTERRORISM GRANTS STATE ALLOCATIONS**

- **FACT SHEET: DEPARTMENT OF HOMELAND SECURITY APPROPRIATIONS ACT OF 2005**

- **FACT SHEET: U.S. DEPARTMENT OF HOMELAND SECURITY 2004 YEAR END REVIEW**

- **FIRST RESPONDER GRANTS**

- **DEPARTMENT OF HOMELAND SECURITY PROGRAMS AND THIS COMPLIANCE SUPPLEMENT**

- **2006 BUDGET REQUEST SHOWS AN INCREASE OF SEVEN PERCENT**

LABORATORIES AND RESEARCH FACILITIES
Homeland Security Centers of Excellence

Courtesy of the United States Government

The Department of Homeland Security is harnessing the nation's scientific knowledge and technological expertise to protect America and our way of life from terrorism. The Department's Science and Technology directorate, through its Office of University Programs, is furthering this mission by engaging the academic community to create learning and research environments in areas critical to Homeland Security.

Through the Homeland Security Centers of Excellence program, Homeland Security is investing in university-based partnerships to develop centers of multi-disciplinary research where important fields of inquiry can be analyzed and best practices developed, debated, and shared.

The Department's Homeland Security Centers of Excellence (HS-Centers) bring together the nation's best experts and focus its most talented researchers on a variety of threats that include agricultural, chemical, biological, nuclear and radiological, explosive and cyber terrorism as well as the behavioral aspects of terrorism.

Homeland Security Centers

Johns Hopkins University (JHU) and its partners have been awarded $15 million over the next three years for the Center for the Study of High Consequence Event Preparedness and Response. This fifth Homeland Security Center of Excellence will study deterrence, prevention, preparedness and response, including issues such as risk assessment, decision-making, infrastructure integrity, surge capacity and sensor networks. In particular, it will study interactions of networks and the need to use models and simulations. (Awarded December 2005)

The Department selected the University of Southern California (partnering with the University of Wisconsin at Madison, New York University, North Carolina State University, Carnegie Mellon University, Cornell University, and others) to house the first HS-Center, known as the Homeland Security Center for Risk and Economic Analysis of Terrorism Events (CREATE). The Department is providing the University of Southern California and its partners with $12 million over the course of the next three years for the study of risk analysis related to the economic consequences of terrorist threats and events. (Awarded November 2003.)

Texas A&M University and its partners have been awarded $18 million over the course of the next three years for the Homeland Security National

Center for Foreign Animal and Zoonotic Disease Defense. Texas A&M University has assembled a team of experts from across the country, which includes partnerships with the University of Texas Medical Branch, University of California at Davis, University of Southern California and University of Maryland. Texas A&M University's HS-Center will work closely with partners in academia, industry and government to address potential threats to animal agriculture including foot and mouth disease, Rift Valley fever, Avian influenza and Brucellosis. Their research on foot and mouth disease will be carried out in close collaboration with Homeland Security's Plum Island Animal Disease Center. (Awarded April 2004.)

The University of Minnesota and its partners have been awarded $15 million over the course of the next three years for the Homeland Security Center for Food Protection and Defense, which will address agro-security issues related to post-harvest food protection. The University of Minnesota's team includes partnerships with major food companies as well as other universities including Michigan State University, University of Wisconsin at Madison, North Dakota State University, Georgia Institute of Technology, Rutgers University, Harvard University, University of Tennessee, Cornell University, Purdue University and North Carolina State University. (Awarded April 2004.)

The University of Maryland and its major partners, the University of California at Los Angeles, the University of Colorado, the Monterey Institute of International Studies, the University of Pennsylvania, and the University of South Carolina, were awarded $12 million over the course of three years for the Homeland Security Center of Excellence for Behavioral and Social Research on Terrorism and Counter-Terrorism. This Center will address a set of broad, challenging questions on the causes of terrorism and strategies to counter terrorism, developing the tools necessary to improve our understanding of, and response to, the magnitude of the threat, examining the psychological impact of terrorism on society, and strengthening the population's resilience in the face of the terrorism. (Awarded January 2005.)

Michigan State University was awarded $10 million over the next five years to house the Center for Advancing Microbial Risk Assessment (CAMRA), jointly funded by the U.S. Department of Homeland Security and the U.S. Environmental Protection Agency. will provide policy-makers and first responders with the information they need to protect human life from biological threats and to set decontamination goals by focusing on two primary objectives. The first objective is a technical mission to develop models, tools, and information that can be used to reduce or eliminate health impacts from the deliberate indoor or outdoor use of biological agents. The second objective is a knowledge management mission to build a national network for information transfer about microbial risk assessment among universities, professionals, and communities. (Awarded October 2005.)

HOMELAND SECURITY URGES SMALL AND MEDIUM-SIZED BUSINESSES TO TAKE STEPS TO PREPARE FOR EMERGENCIES

Courtesy of the United States Government

For Immediate Release
Office of the Press Secretary
Contact: 202-282-8010
October 31, 2005

Ready Business, an extension of the U.S. Department of Homeland Security's successful Ready campaign, is designed to educate owners and managers of small to medium-sized businesses about preparing their employees, operations and assets in the event of an emergency. To help spread this critical message, Homeland Security, in partnership with The Advertising Council, has sponsored new business-specific public service announcements (PSAs). The PSAs will launch nationally this November, and will focus on the affordability and ease of business continuity planning, and the resources available to aid businesses in the process.

"Small businesses form the backbone of our nation's economy, and their emergency preparedness is crucial to keeping our nation secure," said Homeland Security Secretary Michael Chertoff. "A commitment to planning today can protect a business and give it a better chance for survival in the event of a natural disaster, terrorist attack or other emergency."

According to the U.S. Small Business Administration, small businesses represent more than 99 percent of all employers; provide approximately 75 percent of the net new jobs added to the economy; and represent 97 percent of all U.S. exporters. Unfortunately, small to medium-sized businesses are also the most vulnerable in the event of an emergency. By taking some steps ahead of time, many of these businesses can be better prepared to survive and recover after an emergency, thus making the nation and the economy more stable.

Although most businesses agree emergency preparedness is important, too few are taking the necessary steps to prepare. According to an October

2005 survey of small businesses conducted by The Ad Council, 92 percent of respondents said that it is very important or somewhat important for businesses to take steps to prepare for a catastrophic disaster, such as an earthquake, hurricane or terrorist attack. However, only 39 percent said that their company has a plan in place in the event of such a disaster. Qualitative research with this audience demonstrated that even though many acknowledge the value of preparedness, they see time, workforce and money constraints prohibiting them from developing a business continuity plan. The goal of these new Ready Business ads is to show that developing an emergency preparedness plan can be relatively easy and cost-affordable.

"The Ready Business campaign has been very well received during the past year, and many businesses have been motivated to take steps toward preparedness," said Ad Council President Peggy Conlon. "I believe this new round of work will resonate with the business community on a personal level, and reach those businesses that still have not taken advantage of the important resources available to them."

The new PSA campaign, titled "Experience Lines," was created by Chicago-based business-to-business agency Slack Barshinger and includes print, radio and Internet PSAs. Designed to reach managers and owners of small to medium-sized businesses in a number of industries, the advertisements encourage business people to visit the Ready Business campaign website at http://www.ready.gov . The website and other Ready Business materials provide practical steps and easy-to-use templates that can assist businesses in tasks such as creating an evacuation plan; planning for fire safety; considering people with special needs in their plans; securing facilities and equipment; and reviewing insurance coverage.

The U.S. Department of Homeland Security and The Advertising Council launched the Ready Business campaign in September 2004. This extension of Homeland Security's successful Ready campaign, which has helped millions of individuals and families prepare for emergencies, focuses on business preparedness. Ready Business helps owners and managers of small- to medium-sized businesses prepare their employees, operations and assets in the event of an emergency. Visit http://www.ready.gov for more information.

Homeland Security

Office for Domestic Preparedness Grants
2004 COUNTERTERRORISM GRANTS STATE ALLOCATIONS

STATE NAME	STATE GRANTS ALLOCATION TOTAL	PREVENTION & DETERRENCE ALLOCATION TOTAL	CITIZEN CORPS ALLOCATION TOTAL	TOTALS
ALABAMA	27,972,000	8,300,000	581,000	36,853,000
ALASKA	14,774,000	4,384,000	307,000	19,465,000
ARIZONA	31,304,000	9,289,000	650,000	41,243,000
ARKANSAS	21,871,000	6,490,000	454,000	28,815,000
CALIFORNIA	133,174,000	39,517,000	2,766,000	175,457,000
COLORADO	28,041,000	8,321,000	582,000	36,944,000
CONNECTICUT	24,448,000	7,255,000	508,000	32,211,000
DELAWARE	15,336,000	4,551,000	319,000	20,206,000
DISTRICT OF COLUMBIA	14,524,000	4,310,000	302,000	19,136,000
FLORIDA	69,967,000	20,762,000	1,453,000	92,182,000
GEORGIA	41,964,000	12,452,000	872,000	55,288,000
HAWAII	16,839,000	4,997,000	350,000	22,186,000
IDAHO	17,169,000	5,095,000	357,000	22,621,000
ILLINOIS	55,841,000	16,570,000	1,160,000	73,571,000
INDIANA	33,717,000	10,005,000	700,000	44,422,000
IOWA	22,650,000	6,721,000	470,000	29,841,000
KANSAS	21,891,000	6,496,000	455,000	28,842,000
KENTUCKY	26,621,000	7,899,000	553,000	35,073,000
LOUISIANA	27,959,000	8,296,000	581,000	36,836,000
MAINE	17,009,000	5,047,000	353,000	22,409,000
MARYLAND	31,310,000	9,291,000	650,000	41,251,000
MASSACHUSETTS	34,640,000	10,279,000	719,500	45,638,500
MICHIGAN	47,083,000	13,971,000	978,000	62,032,000
MINNESOTA	29,804,000	8,844,000	619,000	39,267,000
MISSISSIPPI	22,426,000	6,655,000	466,000	29,547,000
MISSOURI	32,046,000	9,509,000	666,000	42,221,000
MONTANA	15,687,000	4,655,000	326,000	20,668,000
NEBRASKA	18,502,000	5,490,000	384,000	24,376,000
NEVADA	20,028,000	5,943,000	416,000	26,387,000
NEW HAMPSHIRE	16,942,000	5,027,000	352,000	22,321,000
NEW JERSEY	42,067,000	12,483,000	874,000	55,424,000
NEW MEXICO	18,934,000	5,619,000	393,000	24,946,000
NEW YORK	78,362,000	23,253,000	1,628,000	103,243,000
NORTH CAROLINA	41,140,000	12,208,000	855,000	54,203,000
NORTH DAKOTA	14,741,000	4,374,000	306,000	19,421,000
OHIO	51,791,000	15,368,000	1,076,000	68,235,000
OKLAHOMA	24,563,000	7,289,000	510,000	32,362,000
OREGON	24,658,000	7,317,000	512,000	32,487,000
PENNSYLVANIA	54,929,500	16,300,000	1,141,000	72,370,500
RHODE ISLAND	16,237,000	4,818,000	337,000	21,392,000
SOUTH CAROLINA	26,670,000	7,914,000	554,000	35,138,000
SOUTH DAKOTA	15,177,000	4,504,000	315,000	19,996,000
TENNESSEE	32,475,000	9,636,000	675,000	42,786,000
TEXAS	87,369,000	25,926,000	1,815,000	115,110,000
UTAH	20,518,000	6,089,000	426,000	27,033,000
VERMONT	14,681,000	4,356,000	305,000	19,342,000
VIRGINIA	37,614,000	11,161,000	781,000	49,556,000
WASHINGTON	33,408,000	9,913,000	694,000	44,015,000
WEST VIRGINIA	18,752,000	5,564,000	389,000	24,705,000
WISCONSIN	31,251,000	9,273,000	649,000	41,173,000
WYOMING	14,276,000	4,236,000	297,000	18,809,000
PUERTO RICO	25,817,000	7,661,000	536,000	34,014,000
VIRGIN ISLANDS	4,561,000	1,353,000	95,000	6,009,000
AMERICAN SAMOA	4,384,000	1,301,000	91,000	5,776,000
GUAM	4,719,000	1,400,000	98,000	6,217,000
NORTHERN MARIANA ISLANDS	4,425,000	1,313,000	92,000	5,830,000
TOTAL	**1,675,058,500**	**497,050,000**	**34,793,500**	**2,206,902,000**

U.S. Department of Homeland Security

April 1, 2004

**Homeland
Security**

Office for Domestic Preparedness Grants
2004 URBAN AREA SECURITY INITIATIVE (UASI) ALLOCATIONS

New York, NY	$46,729,722.00
Chicago, IL	$33,940,783.00
Washington/NCR, DC	$29,128,623.00
Los Angeles, CA	$28,101,720.00
San Francisco, CA	$26,325,036.00
Philadelphia, PA	$22,942,595.00
Houston, TX	$19,837,747.00
Miami, FL	$19,033,677.00
Boston, MA	$19,018,846.00
Jersey City, NJ	$17,011,348.00
Seattle, WA	$16,418,562.00
Baltimore, MD	$15,824,825.00
Santa Ana, CA	$14,969,683.00
Newark, NJ	$14,965,282.00
Detroit, MI	$13,673,445.00
Kansas City, MO	$13,217,201.00
Cincinnati, OH	$12,676,037.00
Minneapolis, MN	$12,207,883.00
Phoenix, AZ	$12,128,223.00
Dallas, TX	$12,126,689.00
Long Beach, CA	$12,064,488.00
Pittsburgh, PA	$11,907,806.00
St. Louis, MO	$10,721,421.00
Atlanta, GA	$10,680,857.00
Las Vegas, NV	$10,468,892.00
San Diego, CA	$10,418,116.00
Cleveland, OH	$10,398,748.00
Anaheim, CA	$10,284,651.00
Milwaukee, WI	$10,117,948.00
Indianapolis, IN	$10,091,984.00
Buffalo, NY	$10,036,290.00
Memphis, TN	$10,008,079.00
San Jose, CA	$9,923,545.00
New Haven, CT	$9,576,127.00
Tampa, FL	$9,220,635.00
Louisville, KY	$8,934,634.00
Orlando, FL	$8,713,496.00
Columbus, OH	$8,656,169.00
Denver, CO	$8,595,347.00
Portland, OR	$8,112,992.00
Sacramento, CA	$7,977,579.00
Oakland, CA	$7,808,348.00
St. Paul, MN	$7,781,726.00
Charlotte, NC	$7,361,266.00
Baton Rouge, LA	$7,151,362.00
New Orleans, LA	$7,110,625.00
Fresno, CA	$7,034,646.00
Albany, NY	$6,813,046.00
Richmond, VA	$6,504,772.00
San Antonio, TX	$6,263,976.00
Total UASI Grants to Urban Areas	**$671,017,498.00**

U.S. Department of Homeland Security   April 1, 2004

Office for Domestic Preparedness Grants
2004 UASI TRANSIT SECURITY GRANTS PROGRAM

NY - MTA - NYC Transit	$9,941,000.00
Chicago - METRA	$3,017,064.00
Washington, DC - WMATA	$2,792,738.00
New York - LIRR	$2,715,984.00
New York - Metro-North	$2,679,425.00
Chicago Transit Authority	$2,469,404.00
Boston - MBTA	$2,109,601.00
Los Angeles - Metrolink	$1,971,111.00
San Francisco - SF Bay Area Rapid Transit District	$1,612,884.00
Boston, MA - Mass. Transportation Bay Authority	$1,594,971.00
Philadelphia - SEPTA	$1,585,558.00
Philadelphia - SEPTA - subway	$1,521,652.00
Atlanta - Metropolitan Atlanta Rapid Transit Agency	$1,483,046.00
NY - Port Authority of NY/NJ	$1,249,734.00
Maryland - MARC	$1,031,630.00
Indianapolis - NICTD	$795,280.00
San Franciso - Caltrain	$795,280.00
Virginia - VRE	$795,280.00
San Jose CA - ACE	$795,280.00
Los Angeles - LA County Metropolitan Transportation Authority	$795,280.00
Miami - Tri-Rail	$795,280.00
Miami - Miami-Dade Transit Agency	$795,280.00
New Haven - Shoreline East -	$795,280.00
Mass Transit Administration - Maryland DOT	$795,280.00
San Diego - Coaster	$795,280.00
Seattle - Sound Transit	$795,280.00
Phil-NJ- PATCO	$795,280.00
Cleveland - Greater Cleveland Regional Transit Authority	$795,280.00
Dallas - Trinity Railway Express	$795,280.00
NY- MTA - Staten Island Railway	$795,280.00

Total UASI Transit Security Grants   **$49,705,002.00**

FACT SHEET: DEPARTMENT OF HOMELAND SECURITY APPROPRIATIONS ACT OF 2005

Courtesy of the United States Government

October 18, 2004, President George W. Bush signed the FY 2005 Homeland Security Appropriations Act, which provides $28.9 billion in net discretionary spending for the Department of Homeland Security (DHS). This is $1.8 billion more than the FY 2004 enacted level - reflecting a 6.6% increase in funding for the Department over the previous year. Including Project BioShield, mandatory and fee-funded programs, a total of $40.7 billion will be available to the Department in FY 2005.

The funding provided in FY 2005 reflects the ongoing commitment by the Administration and the Congress to secure the homeland. The act will allow the Department to build upon significant investments to date by improving our safeguards against terrorism, while sustaining the many other important departmental activities.

Strengthening Border and Port Security

The Act includes $419.2 million in new funding to enhance border and port security activities, including the expansion of pre-screening cargo containers in high-risk areas and the detection of individuals attempting to illegally enter the United States.

Additional funding for the U.S. Coast Guard (+$500 million, an 8.6-percent increase) will upgrade port security efforts and provide additional resources to implement the Maritime Transportation Security Act. Key enhancements funded by the act include:

The Container Security Initiative (CSI) focuses on pre-screening cargo before it reaches our shores. The act includes an increase of $25 million over the current program funding of $101 million to continue both Phases I and II, as well as to begin the final phase of CSI.

The United States Visitor and Immigrant Status Indicator Technology (US -VISIT) program's first phase was deployed at 115 airports and 14 seaports. US VISIT expedites the arrival and departure of legitimate travelers, while making it more difficult for those intending to do us harm to enter our nation. The act provides $340 million in 2005, an increase of $12 million over the FY 2004 funding, to continue expansion of the US VISIT system.

Aerial Surveillance and Sensor Technology increases the effectiveness of the more than 12,000 Border Patrol agents deployed along the borders, and supports other missions such as drug interdiction. The act includes $64.2 million for CBP to enhance land-based detection and monitoring of movement between the ports. The act also includes $28 million for CBP to increase the flight hours of P-3 aircraft and $12.5 million for long range radar operations.

Radiation Detection Monitors screen passengers and cargo coming into the United States. The act includes $80 million for the next generation of screening devices for our nation's ports of entry.

CBP Targeting Systems aid in identifying high-risk cargo and passengers. The act includes an increase of $20.6 million for staffing and technology acquisition to support the National Targeting Center, trend analysis, and the Automated Targeting Systems.

The Customs Trade Partnership Against Terrorism (C-TPAT) focuses on partnerships to improve security along the entire supply chain, from the factory floor, to foreign vendors, land borders and seaports. The FY 2005 appropriation includes an increase of $15.2 million for this effort.

The act increases the U.S. Coast Guard's budget by 9 percent — from $5.8 billion in FY 2004 to $6.3 billion in FY 2005. In addition to maintaining its ongoing mission, the budget provides over $100 million to support the implementation of the Maritime Transportation Security Act, which will increase the Coast Guard's ability to develop, review and approve vessel and port security plans, improve underwater detection capabilities, and increase the intelligence program. The budget also provides for the Coast Guard's ongoing Integrated Deepwater System initiative, funding the program at $724 million, an increase of $56 million over the FY 2004 funding level.

Enhancing Biodefense

An additional $2.5 billion for Project BioShield will be available starting in FY 2005 for the development and pre-purchase of necessary medical countermeasures against weapons of mass destruction, and improved biosurveillance by expanding air monitoring for biological agents in high-threat cities and high-value targets such as stadiums and transit systems. Specifically, the FY 2005 appropriation funds the following initiatives:

Project BioShield allows the Federal Government to pre-purchase critically needed vaccines and medications for biodefense as soon as experts agree that they are safe and effective enough to be added to the Strategic National Stockpile. The program seeks to encourage the development of necessary medical countermeasures against a biological, radiological, or nuclear attack. Starting in 2005, $2.5 billion will be available for BioShield.

Improving Biosurveillance, within DHS, will involve the Science and Technology (S&T) and Information Analysis and Infrastructure Protection (IAIP) directorates.

In S&T, the act provides a total of $118 million to enhance current environmental monitoring activities. A key component of this initiative will be an expansion and deployment of the next generation of technologies related to the BioWatch Program, a bio-surveillance warning system.

In IAIP, $11 million is appropriated to integrate, in real-time, bio-surveillance data collected from sensors throughout the country and fuse this data with information from health and agricultural surveillance and other terrorist-threat information from the law enforcement and intelligence communities.

National Disaster Medical System (NDMS) is responsible for managing and coordinating the Federal medical response to major emergencies and federally declared disasters. For 2005, the act includes $20 million in FEMA for planning and exercises associated with medical surge capabilities.

Information Analysis and Infrastructure Protection

The act provides $894 million, a 7 percent increase from FY 2004 to Information Analysis and Infrastructure Protection (IAIP), which will enhance capabilities to receive intelligence and information from an expanded set of sources, to assess the vulnerabilities of the nation's assets and critical infrastructure, to assess consequences, and to add capabilities in remediation and protective actions. Key provisions include:

Threat Determination and Assessment provides tools and unique analytical capability to enhance the Government's ability to integrate, synchronize and correlate sources of information relating to homeland security, emanating from both traditional (Intelligence and federal law enforcement communities) and non-traditional (state and local governments and private industry) sources, and integrate that knowledge with an understanding of exploitable infrastructure vulnerabilities.

$67.4 million, a $2.1 million increase over FY 2004, to expand the capabilities of the National Cyber Security Division (NCSD), which implements the public and private sector partnership protecting cyber security as it identifies, analyzes, and reduces threats and vulnerabilities; disseminates threat warning information; and coordinates cyber incident preparedness, response, and recovery efforts.

Improving Aviation Security

$5.1 billion for the Transportation Security Administration, including aviation security fees, a $679 million increase over 2004. These funds will be used to continue to improve the quality and efficiency of screening operations through additional screener training, stronger management controls of screener performance, and technology automation.

The act includes $475 million to continue deploying more efficient baggage screening solutions at our nation's busiest airports. This funding will be used to improve the integration of explosive detection system (EDS) equipment into individual airports' baggage processing. This will increase security effectiveness and promote greater efficiency.

The act includes $115 million for air cargo security, to continue the research and deployment of screening technology started in FY 2004 and to increase air cargo inspectors.

In addition, the Federal Air Marshals (FAMS) program, which has been moved to Immigration and Customs Enforcement (ICE), receives $663 million in the act, an increase of $50 million over the FY 2004 level.

$61 million is appropriated to the DHS Science and Technology directorate, to accelerate development of more effective technologies to counter the threat of portable anti-aircraft missiles.

Support for State and Local Governments and First Responders

The act provides a total of $4 billion for state and local assistance programs.

State-based formula grants are funded at $1.5 billion, including $400 million for law enforcement, with provisions directing the use of the per capita formula. The "all hazards" Emergency Management Performance Grant program is funded at $180 million.

Urban Area Security Initiative (UASI) grants are provided $885 million, below the request of $1.45 billion. The bill provides a total of $315 million in transportation security grants - in particular, $150 million each for port security grants and rail/transit security grants.

Firefighter assistance grants are funded at $715 million, including $65 million for hiring, compared with the request of $500 million. The statement of managers calls for retaining the program's "all hazards" focus.

The act recognizes the Department's implementation of HSPD-8, and sets deadlines for establishing first responder preparedness levels in January 2005, and releasing the National Preparedness Goal in March 2005.

Enhancing Immigration Security and Enforcement

The Act provides an increase of $179 million for improvements in immigration enforcement both domestically and overseas, including $123 million for the detention and removal of illegal aliens. To enhance immigration security and enforcement, the act includes:

Detention and Removal

An increase of $123 million in FY 2005 will expand ongoing fugitive apprehension efforts and the removal from the United States of jailed offenders, support additional detention and removal capacity.

Immigration Enforcement appropriated funding increases by $56 million for detecting and locating individuals in the United States who are in violation of immigration laws, or who are engaging in immigration-related fraud and will improve visa security by working cooperatively with U.S. consular offices to review visa applications.

Eliminating the Immigration Backlog

The act includes $160 million in total resources to continue progress toward a six-month processing for all immigration applications, while maintaining security and continuing the President's multi-year $500 million initiative to reduce the backlog of applications. CIS has continued the focus on quality improvements and expanded national security checks, such as performing background name checks on all applications before approval.

Increasing DHS Preparedness and Response Capacity

The bill includes $3.1 billion for the Emergency Preparedness and Response Directorate. This funding supports the Nation's ability to prepare for, mitigate against, respond to and recover from natural and man made disasters. This includes $2 billion for the Disaster Relief Fund to allow DHS to provide support to states for response and recovery to unforeseen emergencies and natural disasters.

Strengthening the National Incident Management System (NIMS)

The act provides $15 million for the National Incident Management System (NIMS). The NIMS provides a national framework for Federal, State, Territorial, Tribal, and local jurisdictions to work together more effectively to prevent, prepare for, respond to, and recover from all domestic incidents. The NIMS funding will be used to develop NIMS related training, guidance and other publications to support NIMS implementation. The funding will also be used to support effective resource management through the development of a national resource management system, an inventory of federal response assets, and the development of a national credentialing system.

October 18, 2004

FACT SHEET: U.S. DEPARTMENT OF HOMELAND SECURITY 2004 YEAR END REVIEW

Courtesy of the United States Government

The following is a snapshot of 2004 accomplishments and statistics for the U.S. Department of Homeland Security:

Customs and Border Protection (CBP):
- 428 million passengers and pedestrians, including 262 million aliens, were processed at land, air, and sea ports of entry. Of that number over 643,000 aliens were deemed inadmissible under U.S. law.
- 1,158,800 illegal aliens were arrested by Border Patrol agents between our official ports of entry.
- The Container Security Initiative (CSI) was expanded to include 21 countries. CSI is now operational in 33 foreign ports in Europe, Asia, and Africa. The port of Dubai recently became the first Middle Eastern port to participate in CSI.
- Three months ahead of schedule, the Integrated Automated Fingerprint System (IAFIS) is now operational at all Border Patrol stations. From September through November, over 23,000 individuals with criminal records have been identified and arrested. 84 of those detained were murder suspects and 151 were wanted for sexual assault.
- The Customs-Trade Partnership Against Terror (C-TPAT) has become the largest government/private partnership to arise from September 11th. Just over 8,000 private sector members have applied to participate in C-TPAT.
- CBP officers and agents made 56,321 seizures of illegal drugs, with a total weight of 2,199,619 pounds. Of this number, CBP officers at official ports of entry made 47,744 seizures nationwide, weighing 844,222 pounds and worth an estimated $1 billion. CBP Border

Patrol agents made 8,577 seizures, totaling 1,355,397 pounds of illegal drugs worth an estimated $1.62 billion between the official ports of entry.

- Together with Immigration and Customs Enforcement (ICE), CBP seized more than $138 million worth of counterfeit goods in FY 2004, up from $94 million worth of counterfeit goods in FY 2003.

Federal Emergency Management Agency (FEMA):
- FEMA provided $2.25 billion in aid for individuals and families affected by disasters. The outlay included $1.29 billion in housing assistance, $918 million for other needs assistance, such as medical expenses and personal property losses, and $30.98 million in unemployment benefits.
- More than 1.1 million hurricane victims have registered for assistance since mid-August, the highest ever. $1.43 billion has been spent for individual assistance needs and $1.15 billion in public assistance for the state and local governments.
- 15,560 federal workers were engaged in response and recovery operations for the declared disasters of 2004, including more than 11,000 FEMA personnel and 1,900 disaster medical specialists. As part of the massive response effort in Florida and other hard hit states this past fall, 163 million pounds of ice, 10.8 million gallons of water, 14 million meals-ready-to-eat and 151,000 rolls of plastic roofing material were delivered to help meet immediate emergency needs.

Federal Law Enforcement Training Center (FLETC):
- Provided basic and advance law enforcement training to more than 44,750 students, representing 81 federal agencies, as well as state, local and international law enforcement organizations.
- Aggressively pursued new initiatives in support of homeland security; developing counter-terrorism training programs and facilities; enhancing intelligence awareness and analysis training offerings; and incorporating sophisticated technologies, such as computer generated or controlled simulations, into training.

Immigration & Customs Enforcement (ICE):
- ICE was the second-largest federal contributor to the nation's Joint Terrorism Task Forces (JTTFs) with more than 300 ICE agents assigned to JTTFs nationwide.

- More than 2,500 criminal investigations were conducted involving the illegal export of U.S. arms and strategic technology, including Weapons of Mass Destruction.
- ICE made 1,368 arrests and brought 895 indictments for money laundering and other financial crimes, exceeding arrests and indictments of the prior fiscal year. ICE seized more than $202 million worth of currency, bank accounts, properties and vehicles as a result of financial investigations.
- More than 4,600 child sex predators were apprehended nationwide and over 2,100 child sex predators were deported. The first child sex tourism arrests were made under the Protect Act.
- A 112 percent increase over the prior year for fugitive apprehensions resulted in more than 7,200 arrests. More than 150,000 aliens were removed in FY 2004, 53 percent of who were criminals. This is an all-time record.
- Federal Protective Officers were responsible for 4,426 arrests - a 58 percent increase over the previous fiscal year. They responded to 430 bomb threats and 877 calls about suspicious packages and other items at federal facilities.

Transportation Security Administration (TSA):

Passenger screening has been effective in 2004 by keeping 6,501,193 prohibited items from coming on board aircrafts. The following is a partial list of prohibited items found during screening in 2004:

- 1,895,915 Knives
- 3,285,994 Other Cutting Instruments
- 294,694 Clubs
- 20,509 Box Cutters
- 598 Firearms
- 693,548 Incendiaries
- Over 3,000 arrests were made at security checkpoints.
- Approximately 650 million passengers traveled by air in 2004. 1.8 million Passengers traveled per day and experienced an average screening peak wait time under 12 minutes and an average wait time of 3 minutes in 2004.
- Approximately 600 million checked bags were screened using advanced explosive technologies in 2004.

United States Citizenship & Immigration Services (USCIS):
- 500,000 new United States citizens were naturalized.
- 9,000 active duty military personnel were naturalized through expedited processing.
- 35 million background checks of persons petitioning for immigration benefits were conducted.
- More than 20,000 children from around the world were adopted by U.S. families due to petitions processed by USCIS.
- Almost 50 million visitors sought information about immigration benefits and procedures from the USCIS web-site.

United States Coast Guard (USCG):
- 255,233 pounds of cocaine were seized breaking the record set in 1997.
- 10,348 migrants were interdicted.
- 5,498 lives were saved and 30,895 search-and-rescue cases were conducted.
- The Maritime Transportation Security Act (MTSA) was implemented. It is the largest maritime regulatory project in our nation's history, which entailed the establishment of 43 Area Maritime Security Committees as well as the creation of 43 Area Maritime Security Plans, almost 9,200 Vessel Security Plans, and over 3,100 Facility Security Plans.

United States Secret Service (USSS):
- 30 individuals involved in global cyber organized crime, domestically and internationally, were arrested through Operation Firewall. Industry experts estimate that $1 billion in total fraud loss was prevented.

  Completed 13,395 criminal investigations and arrested 5,566 individuals. Of these, 1,956 individuals were arrested for manufacturing or possessing counterfeit U.S. currency, which resulted in the seizure of 499 counterfeit production plants and $46.5 million in counterfeit currency.
- The Secret Service Electronic Crimes Task Force Initiative was expanded to include 15 task forces that work with federal, state and local law enforcement agencies across the country, prosecutors and experts from the private sector and academia.

US-VISIT:

- In January 2004, US-VISIT (was successfully implemented at all 115 U.S. international airports and 14 seaports. Since that time, more than 370 people with records of criminal or immigration violations have been prevents from entering the U.S. by Customs and Border Protection (CBP) Officers.

- On September 30, Visa Waiver Program (VWP) travelers were included in US-VISIT and the program has now processed more than 14 million travelers while not increasing wait times and earning praise for its privacy efforts.

- US-VISIT is now operational at the nation's 50 busiest land border crossings where significant time savings are already occurring.

FIRST RESPONDER GRANTS

Courtesy of the United States Government

These grant programs represent those Office for Domestic Preparedness (ODP) grant programs generally available to state and local agencies.

*For more information on each State Administrative Agency, please call the ODP Centralized Scheduling and Information Desk at 800-368-6498.

GRANT TITLE: State and Local Domestic Preparedness Exercise Support - CFDA Number 97.006

OVERVIEW: The objective is to enhance the capacity of State and local first responders to respond to a weapons of mass destruction (WMD) terrorism incident involving chemical, biological, radiological, nuclear, and explosive devices. Funds will be used to provide support for planning and conducting exercises at the National, State, and local levels.

ELIGIBLE APPLICANTS: Eligible applicants are public or private organizations with the expertise and experience to provide assistance to State and local jurisdictions; to facilitate, conduct, and evaluate exercises; and/or to develop guidance, materials and publications related to the conduct of exercises or identification of lessons learned.

DEADLINE: The deadline for FY 2001 funds is open.

APPLICATION INFORMATION: For more information go to www.ojp.usdoj.gov/odp/ and click grants.

LINKS:

Grants.gov: http://www.grants.gov/

Program website: http://www.ojp.usdoj.gov/odp/grants programs.htm

Full CFDA description: http://12.46.245.173/pls/portal30/CATA-LOG.PROGRAM TEXT RPT.SHOW?p arg names=prog nbr&p arg values=97.006

GRANT TITLE: State and Local Domestic Preparedness Technical Assistance - CFDA Number 97.007

OVERVIEW: The objective is to provide direct assistance to State and local governments to improve their ability to prevent, respond to, and recover from threats of terrorism involving weapons of mass destruction (WMD). The program goals are to enhance the ability of State and local jurisdictions to develop, plan and implement a program for WMD preparedness and to sustain and maintain specialized equipment. Recipients will provide technical assistance, in response to task orders issued by the Office for Domestic Preparedness (ODP), to States and local units of government in developing, planning, and implementing a program for WMD preparedness.

ELIGIBLE APPLICANTS: Applicants may be public or private organizations with the expertise and experience to provide a specialized service or a full range of technical assistance for state and local governments to prevent, respond to, and recover from a WMD terrorism incident.

DEADLINE: The deadline for FY 2004 funding expired July 1, 2004.

APPLICATION INFORMATION: For more information go to www.ojp.usdoj.gov/odp/ and click grants.

LINKS:

Program website: http://www.ojp.usdoj.gov/odp/grants programs.htm

Full CFDA description: http://12.46.245.173/pls/portal30/CATA-LOG.PROGRAM TEXT RPT.SHOW?p arg names=prog nbr&p arg values=97.008

GRANT TITLE: Emergency Management Performance Grants - CFDA Number 97.042

OVERVIEW: The objective is to assist the development, maintenance, and improvement of State and local emergency management capabilities, which are key components of a comprehensive national emergency management system for disasters and emergencies that may result from natural disasters or accidental or man-caused events. By combining former program activities into the Emergency Management Performance Grant (EMPG), DHS is providing States the flexibility to allocate funds according to risk and to address the most urgent State and local needs in disaster mitigation, preparedness, response, and recovery. Working within the standard Federal government grant administration process,

EMPG provides the support that State and local governments need to achieve measurable results in key functional areas of emergency management: 1.) Laws and Authorities; 2.) Hazard Identification and Risk Assessment; 3.) Hazard Management; 4.) Resource Management; 5.) Planning; 6.) Direction, Control, and Coordination; 7.) Communications and Warning; 8.) Operations and Procedures; 9.) Logistics and Facilities; 10.) Training; 11.) Exercises; 12.) Public Education and Information; and 13.) Finance and Administration. EMPG funds may be used for necessary and essential expenses involved in the development, maintenance, and improvement of State and local emergency management programs. EMPG may be used from time to time as the instrument for delivering Federal assistance for specified program activities subject to terms and conditions.

ELIGIBLE APPLICANTS: All States are eligible (including the District of Columbia and territories and possessions of the United States). Local government entities are not eligible to apply directly to DHS.

DEADLINE: All eligible applicants are notified of target dates and any applicable deadlines by way of the annual EMPG guidance package.

APPLICATION INFORMATION: The FY 04 deadline expired.

LINKS:

Program website: http://www.fema.gov/preparedness/empg.shtm

Full CFDA description: http://12.46.245.173/pls/portal30/CATA-LOG.PROGRAM_TEXT_RPT.SHOW?p_arg_names=prog_nbr&p_arg_values=97.042

GRANT TITLE: State Fire Training Systems Grants (National Fire Academy Training Grants) - CFDA Number 97.043

OVERVIEW: To provide financial assistance to State Fire Training Systems for the delivery of a variety of National Fire Academy (NFA) courses/programs. Grant funds are to be used within each State to deliver NFA training courses/programs and for marketing, administrative costs and electronic feedback of student data. Each State shall spend at least $28,000 ($25,000 in FY 02) in costs relating directly to the delivery of course materials. Computer purchases are restricted.

ELIGIBLE APPLICANTS: Representatives from the 50 State Fire Training Systems.

DEPARTMENT OF HOMELAND SECURITY PROGRAMS AND THIS COMPLIANCE SUPPLEMENT

The Department of Homeland Security (DHS) was created through the Homeland Security Act of 2002 on January 23, 2002. On March 1, 2003, several programs, functions, and transactions were transferred from different Federal agencies to DHS. A "cross-walk"of grant and other financial assistance programs from the legacy agency to DHS, with their associated Catalog of Federal Domestic Assistance(CFDA) numbers, is available on this web site under "GRANTS."

Programs identified on the attached Migration Chart were administered by program offices in other Federal agencies: i. e. Federal Emergency Management Agency (FEMA), Department of Health and Human Services (HHS), and the Department of Justice (DOJ). With the transfer of programs to DHS, the agencies'pre-existing CFDA numbers were also transferred. During the first year's transition, legacy agencies'CFDA numbers were retired and new DHS CFDA numbers were assigned; but some grant awards remain active under the old CFDA numbers. Thus, some programs may have dual CFDA numbers and should be audited accordingly.

The diagram chart will assist auditors in identifying and clarifying "legacy"(pre-DHS) programs by CFDA numbers, and the migration, consolidation, and incorporation of those programs into the CFDA 97.067 Homeland Security Grant Program. The chart focuses only on programs administered by the Office for Domestic Preparedness (ODP). For additional clarification, the following definitions will apply to the chart:•"Consolidation"means administrative steps taken to fold or absorb formerly distinct and separate programs into a single program. "Consolidated"program lose their distinct-separate identity.

"Incorporation"means administrative steps taken to move formerly distinct and separate programs under a single program. "Incorporated"programs retain their distinct-separate, identity.•"Fiscal Year"means the Federal Fiscal year which runs from Oct 1 to September 30. Because most States and local jurisdictions have different fiscal year periods than the Federal government, care must be taken to ascertain that the appropriate ODP program, fiscal year(s) and program guidance is being used to audit. There may have been changes to the requirements for each program that has occurred between one Federal fiscal year to the next. In order to avoid confusion, please review the diagram chart and the ODP program guidance link contained in this web site.

"Migration"means the transfer of a program from one CFDA number/program description to another CFDA number/program description.

CONTACT PHONE NUMBER FOR ODP PROGRAM AUDIT-RELATED QUESTIONS: 800-368-6498 (8:00 a.m. -7:00 p.m. Eastern time, Mon. -Fri.)

2006 BUDGET REQUEST INCLUDES A SEVEN PERCENT INCREASE

Courtesy of the United States Government

President George W. Bush's FY 2006 budget request includes a total of $41.1 billion for the Department of Homeland Security. This is an increase of seven percent over the enacted FY 2005 funding, excluding Project BioShield. This year's request demonstrates the Administration's continued commitment to making further improvements to the nation's homeland security.

DHS has made great strides since it was established in March 2003 in integrating the 22 distinct agencies and bureaus, each with its own employees, mission and culture into a single, unified Department whose mission is to secure the homeland. This year's budget request includes several key initiatives that will allow the Department to integrate and consolidate existing security functions to more effectively serve our overall mission and make America safer.

Overall FY 2006 Budget Highlights

Among the entities with significant budgetary increases are Immigration and Customs Enforcement with a 13.5 percent increase and the U.S. Coast Guard with an increase of more than nine percent.

The budget includes the establishment of the Domestic Nuclear Detection Office (DNDO). The DNDO will develop, acquire and support the deployment and improvement of a domestic system to detect and report attempts to import, assemble, or transport a nuclear explosive device, fissile material or radiological material intended for illicit use. The DNDO will be located within DHS and will be jointly staffed with representatives from DHS, the Department of Energy, the Department of Defense, and the Federal Bureau of Investigation (FBI), with coordination between the Department of Justice, Department of State, intelligence community, and other departments as needed.

The budget proposes to consolidate the various DHS screening activities with the formation of the Office of Screening Coordination and Operations (SCO) within the Border and Transportation Security (BTS) directorate. This new organization would enhance terrorist-related screening through comprehensive, coordinated procedures that detect, identify, track and interdict people, cargo and other entities and objects that pose a threat to homeland security. This effort to integrate existing resources to work more efficiently, brings together several similar ongoing screening efforts under one office, including: United States-Visitor and Immigrant Status Indicator Technology (US-VISIT); Secure Flight and Crew Vetting; Free and Secure Trade (FAST); NEXUS/Secure Electronic Network for Travelers Rapid Inspection (SENTRI); Transportation Worker Identification Credential (TWIC); Registered Traveler; Hazardous Materials Trucker Background Checks; and Alien Flight School Checks.

The effectiveness of state and local homeland security assistance can be increased through an approach that closes the most critical gaps in terrorism prevention and preparedness capabilities. Over $2 billion in grants for states and urban areas would be based on assessments of risk and vulnerability, as well as the needs and priorities identified in state and regional homeland security plans. The proposed Targeted Infrastructure Protection program would provide $600 million in integrated grants, enabling DHS to supplement state, local and regional government efforts in their protection of critical national infrastructures such as seaports, mass transit, railways, and energy facilities.

In FY 2006, DHS seeks to consolidate the research, development, test and evaluation (RDT&E) activities within the DHS Science and Technology (S&T) directorate. This consolidation, in the amount of $127 million, will bring the scientific and engineering personnel and other RDT&E resources of the Department under a single accountable authority.

The Department requests $49.9 million to begin to establish a regional structure for DHS and integrate and identify efficiencies within information technology, facilities, and operations centers across DHS. Of the 22 agencies that were brought together to form the Department, twelve have regional and field structures ranging in size from three to thirty offices distributed throughout the nation.

Aviation security is a shared responsibility of the federal government, airports, airlines and traveling public. Airport screening, one element of aviation security, benefits passengers and air carriers by protecting them from threats. These costs should be borne primarily by the beneficiaries of screening services. The budget proposes raising the fee on a typical one-leg ticket from $2.50 one way to $5.50. For passengers traveling multiple legs on a one-way trip, that fee would increase from the current maximum of $5.00 to $8.00. Fees cover nearly the full cost of aviation screening operations.

The President remains committed to ensure America welcomes the contributions of immigrants. The budget continues funding for the President's multi-year $540 million initiative enabling U.S. Citizenship and Immigration Services to reduce the backlog of applications and ensure a six-month processing standard for all applications by the end of 2006.

The budget revolves around five major themes: Revolutionizing the Borders; Strengthening Law Enforcement; Improving National Preparedness and Response; Leveraging Technology; and Creating a 21st Century Department.

Revolutionizing the Borders

Weapons of Mass Destruction (WMD) Detection Technology is an integral part of the DNDO comprehensive strategy to address the threat of nuclear and radiological terrorism. The budget includes $125 million to purchase additional Radiation Portal Monitors (RPMs) and pilot advanced next generation RPMs to detect both gamma and neutron radiation at our borders.

The Container Security Initiative (CSI), which focuses on pre-screening cargo before it reaches our shores, will have a preventative and deterrence effect on the use of global containerized shipping of WMD and other terrorist equipment. Egypt, Chile, India, Philippines, Venezuela, Bahamas and Honduras have been identified as pilots for screening in FY 2006. An increase of $5.4 million over FY 2005 is included in CBP's budget for CSI, for a total request of $138.8 million.

CBP Targeting Systems aid in identifying high-risk cargo and passengers. The budget includes a total of $28.3 million for these system initiatives, which includes a $5.4 million increase.

America's Shield Initiative (ASI) enhances electronic surveillance capabilities along the northern and southern land borders of the U.S. by improving the sensor and video surveillance equipment deployed to guard against the entry of illegal aliens, terrorists, WMDs and contraband into the U.S. The budget includes $51.1 million, an increase of $19.8 million.

US-VISIT, which is proposed for consolidation within the SCO, increases from $340 million to $390 million. The increase will provide for the accelerated deployment of US-VISIT at the land borders and enhance access for border personnel to immigration, criminal and terrorist information.

The Customs Trade Partnership Against Terrorism (C-TPAT) focuses on partnerships all along the entire supply chain, from the factory floor, to foreign vendors, to land borders and seaports. The budget includes an increase of $8.2 million, for a total amount of $54.3 million. The increase will enhance our ability to conduct additional supply chain security validations.

Border Patrol Staffing would increase along the southwest border and coastal areas, in part to replace some Border Patrol agents shifted to the northern border as required by the Patriot Act. An increase of 210 agents and $36.9 million is included in the budget for the Border Patrol. This increases the Border Patrol Agents to 10,949.

Long Range Radar technology is used by the Office of Air and Marine Operations to detect and intercept aircraft attempting to avoid detection while entering the U.S. CBP and the Department of Defense will assume responsibility for operating and maintaining these systems from the FAA beginning in FY 2006. CBP's share is $44.2 million in the budget.

Strengthening Law Enforcement

The Armed Helicopter for Homeland Security Project increases by $17.4 million in the budget. These funds will provide equipment and aircraft modifications to establish armed helicopter capability at five USCG Air Stations. This will provide the USCG and DHS with the tools needed to respond quickly and forcefully to emergency maritime threats. A total of $19.9 million is included in the budget for this project.

The Integrated Deepwater System increases by $242 million to a total of $966 million in FY 2006 to continue the acquisition of the USCG's Maritime Security Cutter-Large, complete design of the Maritime Security Cutter-Medium, promote completion of the Multi-Mission Cutter Helicopter (re-engineered and electronically upgraded HH-65 helicopter) and significantly improve fixed and rotary wing aircraft capabilities. These upgrades will increase awareness and are crucial for an integrated, interoperable border and port security system.

The Response Boat-Medium Project increases the effort to replace the USCG's 41-foot utility boats and other large non-standard boats with assets more capable of meeting all of the USCG's multi-mission operational requirements by $10 million. A total of $22 million is proposed in the budget for this effort.

The Federal Air Marshal Service (FAMS) seeks a total of $688.9 million. This funding will allow ICE to protect air security and promote public confidence in our civil aviation system.

Detention and Removal within ICE increases by $176 million for detention and removal activities. Total increases for this program are approximately 19 percent above the FY 2005.

Temporary Worker Worksite Enforcement increases will more than double the resources available for worksite enforcement including employer audits, investigations of possible violations and criminal case presentations. An increase of $18 million is in the budget.

Federal Flight Deck Officers (FFDO)/Crew Member Self-Defense (CMSD) Training is increased by $11 million in FY 2006 for a total of $36.3 million. This allows for the expansion of the semi-annual firearm re-qualification program for FFDO personnel and to fund the first full year of the CMSD training program.

Improving National Preparedness and Response

Federal assistance for our nation's first responder community. The budget includes $3.6 billion for grants, training, and technical assistance administered by the Office of State and Local Government Coordination and Preparedness (SLGCP). This funding will support state and local agencies as they equip, train, exercise, and assess preparedness for emergencies regardless of scale or cause.

Enhanced Catastrophic Disaster Planning is budgeted at $20 million for FEMA to work with states and localities, as well as other federal agencies, to develop and implement plans that will improve the ability to respond to and to recover from catastrophic disasters.

The Office of Interoperability and Compatibility (OIC) within the S&T Directorate will allow the Department to expand its leadership role in interoperable communications that could be used by every first responder agency in the country. The OIC has currently identified three program areas: communications, equipment, and training. With $20.5 million in FY 2006, the OIC will plan and begin to establish the training and equipment programs, as well as continue existing communication interoperability efforts through the SAFECOM Program.

Replacement of the USCG's High Frequency (HF) Communications System, funded at $10 million in the budget, will replace unserviceable, shore-side, high power high frequency transmitters, significantly improving long-range maritime safety and security communications.

The Rescue 21 project is funded at $101 million in the budget to continue recapitalizing the Coast Guard's coastal zone communications network. This funding will complete system infrastructure and network installations in 14 regions and begin development of regional designs for the remaining 11 regions.

Leveraging Technology

Low Volatility Agent Warning System is a new FY 2006 initiative totaling $20 million. Funding is included to develop a system that will serve as the basis for a warning and identification capability against a set of chemical agents whose vapor pressure is too low to be detected by conventional measures.

Counter-MAN Portable Air Defense Systems (C-MANPADS) funding is increased by $49 million to a total of $110 million in the budget. This program will continue to research the viability of technical countermeasures for commercial aircraft against the threat of shoulder-fired missiles.

Cyber Security is enhanced in the budget to augment a 24/7 cyber threat watch, warning, and response capability that would identify emerging threats and vulnerabilities and coordinate responses to major cyber security incidents. An increase of $5 million is proposed in the budget for this effort, bringing the program total to $73.3 million.

Secure Flight/Crew Vetting requests an increase of $49 million to field the system developed and tested in FY 2005. The funds will support testing, information systems, connectivity to airlines and screen systems and daily operations. This also includes an increase of $3.3 million for crew vetting.

The budget includes $174 million to complete installation of High Speed Operational Connectivity (Hi-SOC) to passenger and baggage screening checkpoints to improve management of screening system performance.

Emerging Checkpoint Technology is enhanced by $43.7 million in FY 2006 to direct additional resources to improve checkpoint explosives screening. This assures that TSA is on the cutting edge, ahead of the development of increasingly well-disguised prohibited items. This proposed increase will result in investing more than $100 million invested in FY 2005 and FY 2006 for new technology to ensure improved screening of all higher risk passengers.

Homeland Secure Data Network (HSDN) includes $37 million in the budget. These funds will streamline and modernize the classified data capabilities in order to facilitate high quality and high value classified data communication and collaboration.

The Homeland Security Operations Center (HSOC) funding is increased by $26.3 million bringing its FY 2006 funded level to $61.1 million. This includes an increase of $13.4 million for the Homeland Security Information Network (HSIN) and an increase of $12.9 million to enhance HSOC systems and operations.

Creating a 21st Century Department

Electronically Managing enterprise resources for government effectiveness and efficiency (eMerge2) funding of $30 million in the budget to continue implementation of a DHS-wide solution that delivers accurate, relevant and timely resource management information to decision makers. By delivering access to critical information across all components, the Department will be able to better support its many front-line activities.

MAX HR funding of $53 million is to continue the design and deployment of a new human resources system. As outlined in final regulations, issued jointly on February 1, 2005, by Secretary Ridge and the Director of the Office of Personnel Management (OPM) Kay Coles James, the MAXHR system provides greater flexibility and accountability in the way employees are paid, developed, evaluated, afforded due process and represented by labor organizations. The goal is a 21st century personnel system that enhances mission-essential flexibility and preserves core civil service principles and the merit system.

The Information Sharing and Collaboration (ISC) program will affect the policy, procedures, technical, process, cultural, and organizational aspects of information sharing and collaboration, including coordinating ISC policy with other federal agencies, drafting technical and operational needs statements, performing policy assessments and analyzing new requirements. The total funding for FY 2006 will be $16.5 million.

Related Information
Budget in Brief Fiscal Year 2006 (PDF, 108 pages - 2.1 MB)
February 7, 2005

Chapter 4   Appendix

**CYBERSPACE THREATS
AND VULNERABILITIES**

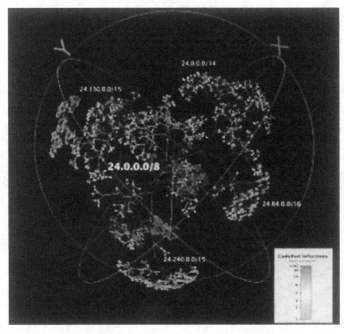

A Mapping of Code Red Penetration on a Portion of the Internet.

Image courtesy UCSD/CAIDA (www.caida.org) © 2002 The Regents of the University of California.

Cyberspace Threats and Vulnerabilities

A Case for Action

The terrorist attacks against the United States that took place on September 11, 2001, had a profound impact on our Nation. The federal government and society as a whole have been forced to reexamine conceptions of security on our home soil, with many understanding only for the first time the lengths to which self-designated enemies of our country are willing to go to inflict debilitating damage.

We must move forward with the understanding that there are enemies who seek to inflict damage on our way of life. They are ready to attack us on our own soil, and they have shown a willingness to use unconventional means to execute those attacks. While the attacks of

September 11 were physical attacks, we are facing increasing threats from hostile adversaries in the realm of cyberspace as well.

A Nation Now Fully Dependent on Cyberspace

For the United States, the information technology revolution quietly changed the way business and government operate. Without a great deal of thought about security, the Nation shifted the control of essential processes in manufacturing, utilities, banking, and communications to networked computers. As a result, the cost of doing business dropped and productivity skyrocketed. The trend toward greater use of networked systems continues.

CYBERSPACE   THREATS   AND   VULNERABILITIES

By 2003, our economy and national security became fully dependent upon information technology and the information infrastructure. A network of networks directly supports the operation of all sectors of our economy—energy (electric power, oil and gas), transportation (rail, air, merchant marine), finance and banking, information and telecommunications, public health, emergency services, water, chemical, defense industrial base, food, agriculture, and postal and shipping. The reach of these computer networks exceeds the bounds of cyberspace. They also control physical objects such as electrical transformers, trains, pipeline pumps, chemical vats, and radars.

Threats in Cyberspace

A spectrum of malicious actors can and do conduct attacks against our critical information infrastructures. Of primary concern is the threat of organized cyber attacks capable of causing debilitating disruption to our Nation's critical infrastructures, economy, or national security. The required technical sophistication to carry out such an attack is high—and partially explains the lack of a debilitating attack to date. We should not, however, be too sanguine. There have been instances where attackers have exploited vulnerabilities that may be indicative of more destructive capabilities.

Uncertainties exist as to the intent and full technical capabilities of several observed attacks. Enhanced cyber threat analysis is needed to address long-term trends related to threats and vulnerabilities. What is known is that the attack tools and methodologies are becoming widely available, and the technical capability and sophistication of users bent on causing havoc or disruption is improving.

As an example, consider the "NIMDA" ("ADMIN" spelled backwards) attack. Despite the fact that NIMDA did not create a catastrophic disruption to the critical infrastructure, it is a good example of the increased technical sophistication showing up in cyber attacks. It demonstrated that the arsenal of weapons available to organized attackers now contains the capability to learn and adapt to its local environment. NIMDA was an automated cyber attack, a blend of a computer worm and a computer virus. It propagated across the Nation with enormous speed and tried several different ways to infect computer systems it invaded until it gained access and destroyed files. It went from nonexistent to nationwide in an hour, lasted for days, and attacked 86,000 computers.

Speed is also increasing. Consider that two months before NIMDA, a cyber attack called Code Red infected 150,000 computer systems in 14 hours.

Because of the increasing sophistication of computer attack tools, an increasing number of actors are capable of launching nationally significant assaults against our infrastructures and cyberspace. In peacetime America's enemies may conduct espionage on our Government, university research centers, and private companies. They may also seek to prepare for cyber strikes during a confrontation by mapping U.S. information systems, identifying key targets, lacing our infrastructure with back doors and other means of access. In wartime or crisis, adversaries may seek to intimidate the nation's political leaders by attacking critical infrastructures and key economic functions or eroding public confidence in information systems.

Cyber attacks on U.S. information networks can have serious consequences such as disrupting critical operations, causing loss of revenue and intellectual property, or loss of life. Countering such attacks requires the development of robust capabilities where they do not exist today if we are to reduce vulnerabilities and deter those with the capabilities and intent to harm our critical infrastructures.

Cyberspace provides a means for organized attack on our infrastructure from a distance. These attacks require only commodity technology, and enable attackers to obfuscate their identities, locations, and paths of entry. Not only does cyberspace provide the ability to exploit weaknesses in our critical infrastructures, but it also provides a fulcrum for leveraging physical attacks by allowing the possibility of disrupting communications, hindering U.S. defensive or offensive response, or delaying emergency responders who would be essential following a physical attack.

In the last century, geographic isolation helped protect the United States from a direct physical invasion. In cyberspace national boundaries have little meaning. Information flows continuously and seamlessly across political, ethnic, and religious divides. Even the infrastructure that makes up cyberspace—software and hardware— is global in its design and development. Because of the global nature of cyberspace, the vulnerabilities that exist are open to the world and available to anyone, anywhere, with sufficient capability to exploit them.

Reduce Vulnerabilities in the Absence of Known Threats

While the Nation's critical infrastructures must, of course, deal with specific threats as they arise, waiting to learn of an imminent attack before addressing important critical infrastructure vulnerabilities is a risky and unacceptable strategy. Cyber attacks can burst onto the Nation's networks with little or no warning and spread so fast that many victims never have a chance to hear the alarms. Even with forewarning, they likely would not have had the time, knowledge, or tools needed to protect themselves. In some cases creating defenses against these attacks would have taken days.

A key lesson derived from these and other such cyber attacks is that organizations that rely on networked computer systems must take proactive steps to identify and remedy their vulnerabilities, rather than waiting for an attacker to be stopped or until alerted of an impending attack. Vulnerability assessment and remediation activities must be ongoing. An information technology security audit conducted by trained professionals to identify infrastructure vulnerabilities can take months. Subsequently, the process of creating a multi-layered defense and a resilient network to remedy the most serious vulnerabilities could take several additional months. The process must then be regularly repeated.

Threat and Vulnerability: A Five-Level Problem

Managing threat and reducing vulnerability in cyberspace is a particularly complex challenge because of the number and range of different types of users. Cyberspace security requires action on multiple levels and by a diverse group of actors because literally hundreds of millions of devices are interconnected by a network of networks. The problem of cyberspace security can be best addressed on five levels.

Level 1, the Home User/Small Business

Though not a part of a critical infrastructure the computers of home users can become part of networks of remotely controlled machines that are then used to attack critical infrastructures. Undefended home and small business computers, particularly those using digital subscriber line (DSL) or cable connections, are vulnerable to attackers who can employ the use of those machines without the owner's knowledge. Groups of such "zombie" machines can then be used by third-party actors to launch denial-of-service (DoS) attacks on key Internet nodes and other important enterprises or critical infrastructures.

Level 2, Large Enterprises

Large-scale enterprises (corporations, government agencies, and universities) are common targets for cyber attacks. Many such enterprises are part of critical infrastructures. Enterprises require clearly articulated, active

CYBERSPACE THREATS AND VULNERABILITIES

information security policies and programs to audit compliance with cybersecurity best practices. According to the U.S. intelligence community, American networks will be increasingly targeted by malicious actors both for the data and the power they possess.

Level 3, Critical Sectors/Infrastructures

When organizations in sectors of the economy, government, or academia unite to address common cybersecurity problems, they can often reduce the burden on individual enterprises. Such collaboration often produces shared institutions and mechanisms, which, in turn, could have cyber vulnerabilities whose exploitation could directly affect the operations of member enterprises and the sector as a whole. Enterprises can also reduce cyber risks by participating in groups that develop best practices, evaluate technological offerings, certify products and services, and share information.

Several sectors have formed Information Sharing and Analysis Centers (ISACs) to monitor for cyber attacks directed against their respective infrastructures. ISACs are also a vehicle for sharing information about attack trends, vulnerabilities, and best practices.

Level 4, National Issues and Vulnerabilities

Some cybersecurity problems have national implications and cannot be solved by individual enterprises or infrastructure sectors alone. All sectors share the Internet. Accordingly, they are all at risk if its mechanisms (e.g., protocols and routers) are not secure. Weaknesses in widely used software and hardware products can also create problems at the national level, requiring coordinated activities for the research and development of improved technologies. Additionally, the lack of trained and certified cybersecurity professionals also merits national-level concern.

Level 5, Global

The worldwide web is a planetary information grid of systems. Internationally shared standards enable interoperability among the world's computer systems. This interconnectedness, however, also means that problems on one continent have the potential to affect computers on another. We therefore rely on international cooperation to share information related to cyber issues and, further, to prosecute cyber criminals. Without such cooperation, our collective ability to detect, deter, and minimize the effects of cyber-based attacks would be greatly diminished.

New Vulnerabilities Requiring Continuous Response

New vulnerabilities are created or discovered regularly. The process of securing networks and systems, therefore, must also be continuous. The Computer Emergency Response Team/Coordination Center (CERT/CC) notes that not only are the numbers of cyber incidents and attacks increasing at an alarming rate, so too are the numbers of vulnerabilities that an attacker could exploit. Identified computer security vulnerabilities—faults in software and hardware that could permit unauthorized network access or allow an attacker to cause network damage—increased significantly from 2000 to 2002, with the number of vulnerabilities going from 1,090 to 4,129.

The mere installation of a network security device is not a substitute for maintaining and updating a network's defenses. Ninety percent of the participants in a recent Computer Security Institute survey reported using antivirus software on their network systems, yet 85 percent of their systems had been damaged by computer viruses. In the same survey, 89 percent of the respondents had installed computer firewalls, and 60 percent had intrusion detection systems. Nevertheless, 90 percent reported that security breaches had taken place, and 40 percent of their systems had

Roles and Responsibilites in Securing Cyberspace

	Priority 1	Priority 2	Priority 3	Priority 4	Priority 5
	National Cyberspace Security Response System	National Cyberspace Security Threat and Vulnerability Reduction System	National Cyberspace Security Awareness and Training Program	Securing Governments' Cyberspace	National Security and International Cyberspace Security Cooperation
Home User/Small Business		X	X		
Large Enterprises	X	X	X	X	X
Critical Sectors/ Infrastructures	X	X	X	X	X
National Issues and Vulnerabilities	X	X	X	X	
Global					X

been penetrated from outside their network.

The majority of security vulnerabilities can be mitigated through good security practices. As these survey numbers indicate, however, practicing good security includes more than simply installing those devices. It also requires operating them correctly and keeping them current through regular patching and virus updates.

Cybersecurity and Opportunity Cost

For individual companies and the national economy as a whole, improving computer security requires investing attention, time, and money. For fiscal year 2003, President Bush requested that Congress increase funds to secure federal computers by 64 percent. President Bush's investment in securing federal computer networks now will eventually reduce overall expenditures through cost-saving E-Government solutions, modern enterprise management, and by reducing the number of opportunities for waste and fraud.

For the national economy—particularly its information technology industry component—the dearth of trusted, reliable, secure information systems presents a barrier to future growth. Much of the potential for economic growth made possible by the information technology revolution has yet to be realized—deterred in part by cyberspace security risks. Cyberspace vulnerabilities place more than transactions at risk; they jeopardize intellectual property, business operations, infrastructure services, and consumer trust.

Conversely, cybersecurity investments result in more than costly overhead expenditures. They produce a return on investment. Surveys repeatedly show that:

• Although the likelihood of suffering a severe cyber attack is difficult to estimate, the costs associated with a successful one are likely to be greater than the investment in a cybersecurity program to prevent it; and

CYBERSPACE THREATS AND VULNERABILITIES

• Designing strong security protocols into the information systems architecture of an enterprise can reduce its overall operational costs by enabling cost-saving processes, such as remote access and customer or supply-chain interactions, which could not occur in networks lacking appropriate security.

These results suggest that, with greater awareness of the issues, companies can benefit from increasing their levels of cybersecurity. Greater awareness and voluntary efforts are critical components of the *National Strategy to Secure Cyberspace.*

Individual and National Risk Management

Until recently overseas terrorist networks had caused limited damage in the United States. On September 11, 2001, that quickly changed. One estimate places the increase in cost to our economy from attacks to U.S. information systems at 400 percent over four years. While those losses remain relatively limited, that too could change abruptly.

Every day in the United States individual companies, and home computer users, suffer damage from cyber attacks that, to the victims, represent significant losses. Conditions likewise exist for relative measures of damage to occur on a national level, affecting the networks and systems on which the Nation depends:

• Potential adversaries have the intent;

• Tools that support malicious activities are broadly available; and,

• Vulnerabilities of the Nation's systems are many and well known.

No single strategy can completely eliminate cyberspace vulnerabilities and their associated threats. Nevertheless, the Nation must act to manage risk responsibly and to enhance its ability to minimize the damage that results

from attacks that do occur. Through this statement, we reveal nothing to potential foes that they and others do not already know. In 1997 a Presidential Commission identified the risks in a seminal public report. In 2000 the first national plan to address the problem was published. Citing these risks, President Bush issued an Executive Order in 2001, making cybersecurity a priority, and accordingly, increasing funds to secure federal networks. In 2002 the President moved to consolidate and strengthen federal cybersecurity agencies as part of the proposed Department of Homeland Security.

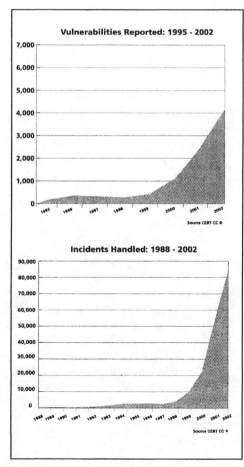

Vulnerabilities Reported: 1995 - 2002

Source CERT CC ©

Incidents Handled: 1988 - 2002

Source CERT CC ©

Government Alone Cannot Secure Cyberspace

Despite increased awareness around the importance of cybersecurity and the measures taken thus far to improve our capabilities, cyber risks continue to underlie our national information networks and the critical systems they manage. Reducing that risk requires an unprecedented, active partnership among diverse components of our country and our global partners.

The federal government could not—and, indeed, should not—secure the computer networks of privately owned banks, energy companies, transportation firms, and other parts of the private sector. The federal government should likewise not intrude into homes and small businesses, into universities, or state and local agencies and departments to create secure computer networks. Each American who depends on cyberspace, the network of information networks, must secure the part that they own or for which they are responsible.

Chapter 5   Appendix

- NATIONAL RESPONSE PLAN

- DEPARTMENT OF HOMELAND SECURITY—WHO BECAME
  PART OF THE DEPARTMENT?

NATIONAL RESPONSE PLAN

Courtesy of the United States Government

The National Response Plan establishes a comprehensive all-hazards approach to enhance the ability of the United States to manage domestic incidents. The plan incorporates best practices and procedures from incident management disciplines-homeland security, emergency management, law enforcement, firefighting, public works, public health, responder and recovery worker health and safety, emergency medical services, and the private sector-and integrates them into a unified structure. It forms the basis of how the federal government coordinates with state, local, and tribal governments and the private sector during incidents. It establishes protocols to help:

- Save lives and protect the health and safety of the public, responders, and recovery workers;
- Ensure security of the homeland;
Prevent an imminent incident, including acts of terrorism, from occurring;
- Protect and restore critical infrastructure and key resources;
- Conduct law enforcement investigations to resolve the incident, apprehend the perpetrators, and collect and preserve evidence for prosecution and/or attribution;
- Protect property and mitigate damages and impacts to individuals, communities, and the environment; and
- Facilitate recovery of individuals, families, businesses, governments, and the environment.

Download the National Response Plan

National Reponse Base Plan and Appendices
http://www.dhs.gov/interweb/assetlibrary/NRPbaseplan.pdf
(PDF, 114 pages, 2MB)

Full Version
http://www.dhs.gov/interweb/assetlibrary/NRP_FullText.pdf (PDF, 426 pages, 4MB) including all annexes, "Emergency Support Function Annexes", "Support Annexes", and "Incident Annexes."

For additional information on the plan or on how to obtain copies of the National Response Plan, please call 800-368-6498.

Training

The Federal Emergency Management Agency (FEMA) Emergency Management Institute offers an online course designed to introduce the National Response Plan to responsders.

FEMA Emergency Management Institute National Response Plan Course:

http://www.training.fema.gov/emiweb/IS/is800.asp

More on the National Response Plan

Local/Federal Response Strategies & Coordination Structure
    http://www.dhs.gov/dhspublic/interapp/editorial/editorial_0569.xml
Prevention, Preparedness, Response & Recovery
    http://www.dhs.gov/dhspublic/interapp/editorial/editorial_0570.xml
Signatory Partners
    http://www.dhs.gov/dhspublic/interapp/editorial/editorial_0572.xml
Fact Sheet: National Response Plan
    http://www.dhs.gov/dhspublic/interapp/press_release/press_release_05
81.xml

Additional Resources

National Response Plan Brochure
    http://www.dhs.gov/interweb/assetlibrary/NRP_Brochure.pdf (PDF, 2
pages 3MB)
National Incident Management System
    http://www.fema.gov/nims/

HISTORY: WHO BECAME PART OF THE DEPARTMENT?

Courtesy of the United States Government

The agencies slated to become part of the Department of Homeland Security will be housed in one of four major directorates: Border and Transportation Security, Emergency Preparedness and Response, Science and Technology, and Information Analysis and Infrastructure Protection.

The Border and Transportation Security directorate will bring the major border security and transportation operations under one roof, including:

- The U.S. Customs Service (Treasury)
- The Immigration and Naturalization Service (part) (Justice)

- The Federal Protective Service
- The Transportation Security Administration (Transportation)
- Federal Law Enforcement Training Center (Treasury)
- Animal and Plant Health Inspection Service (part)(Agriculture)
- Office for Domestic Preparedness (Justice)
- The Emergency Preparedness and Response directorate will oversee domestic disaster preparedness training and coordinate government disaster response. It will bring together:
- The Federal Emergency Management Agency (FEMA)
- Strategic National Stockpile and the National Disaster Medical System (HHS)
- Nuclear Incident Response Team (Energy)
- Domestic Emergency Support Teams (Justice)
- National Domestic Preparedness Office (FBI)

The Science and Technology directorate will seek to utilize all scientific and technological advantages when securing the homeland. The following assets will be part of this effort:

- CBRN Countermeasures Programs (Energy)
- Environmental Measurements Laboratory (Energy)
- National BW Defense Analysis Center (Defense)
- Plum Island Animal Disease Center (Agriculture)

The Information Analysis and Infrastructure Protection directorate will analyze intelligence and information from other agencies (including the CIA, FBI, DIA and NSA) involving threats to homeland security and evaluate vulnerabilities in the nation's infrastructure. It will bring together:

- Federal Computer Incident Response Center (GSA)
- National Communications System (Defense)
- National Infrastructure Protection Center (FBI)
- Energy Security and Assurance Program (Energy)
- The Secret Service and the Coast Guard will also be located in the Department of Homeland Security, remaining intact and reporting directly to the Secretary. In addition, the INS adjudications and benefits programs will report directly to the Deputy Secretary as the U.S. Citizenship and Immigration Services.
- Historical Documents

DHS Organizational Chart

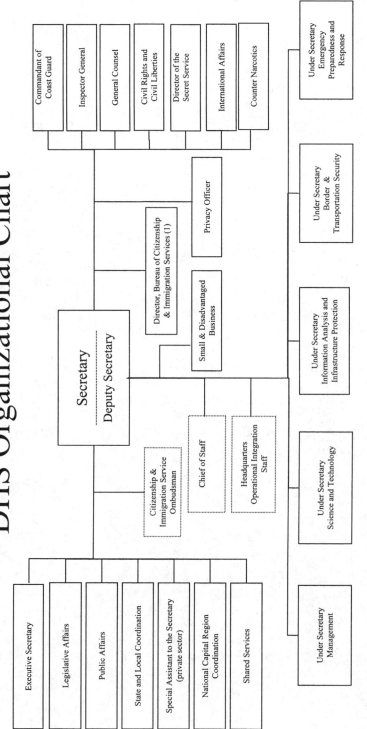

Secretary

Deputy Secretary

Executive Secretary

Legislative Affairs

Public Affairs

State and Local Coordination

Special Assistant to the Secretary (private sector)

National Capital Region Coordination

Shared Services

Citizenship & Immigration Service Ombudsman

Chief of Staff

Headquarters Operational Integration Staff

Director, Bureau of Citizenship & Immigration Services (1)

Privacy Officer

Small & Disadvantaged Business

Commandant of Coast Guard

Inspector General

General Counsel

Civil Rights and Civil Liberties

Director of the Secret Service

International Affairs

Counter Narcotics

Under Secretary Management

Under Secretary Science and Technology

Under Secretary Information Analysis and Infrastructure Protection

Under Secretary Border & Transportation Security

Under Secretary Emergency Preparedness and Response

August 2004

Homeland Security

Chapter 6   Appendix

**NATIONAL STRATEGY
FOR HOMELAND SECURITY**

Foreword

In July 2002 the President approved the *National Strategy for Homeland Security*, a road map for the national effort to prevent and respond to acts of terrorism in the United States. The *National Strategy* recognizes the vital role of state and local public safety agencies in providing for the security of our homeland. In February 2003 the President signed into law the Fiscal Year (FY) 2003 Omnibus Appropriations Act which provides state and local governments with the vital funding they require to participate in the national effort to combat terrorism.

The U.S. Department of Homeland Security (DHS), Office for Domestic Preparedness (ODP) FY 2003 Urban Areas Security Initiative (UASI) reflects a confluence of important Presidential initiatives designed to enhance the preparedness of the nation to combat terrorism. Whereas most states and municipalities have strengthened their overall capability to respond to acts of terrorism involving chemical, biological, radiological, nuclear or explosive (CBRNE) weapons, there continues to be room for improvement in meeting our national priorities of preventing and responding to terrorist attacks.

The Office for Domestic Preparedness is providing financial assistance directly to selected jurisdictions through the Fiscal Year (FY) 2003 Urban Areas Security Initiative. This financial assistance is being provided to address the unique equipment, training, planning and exercise needs of large high threat urban areas, and to assist them in building an enhanced and sustainable capacity to prevent, respond to, and recover from threats or acts of terrorism.

The Department of Homeland Security looks forward to working with all of you in this critical national effort to secure our homeland.

Tom Ridge
Secretary
Department of Homeland Security

Contents

UASI GRANT APPLICATION
**Actual application must be made online through GMS

✍✍ Standard Form 424

✍✍ Assurances

✍✍ Certifications Regarding Lobbying; Debarment, Suspension and Other Responsibility Matters; and Drug-Free Workplace Requirements

✍✍ Non-supplanting Certification

I.   Background

The U.S. Department of Homeland Security (DHS), through the Office for Domestic Preparedness (ODP), is providing financial assistance directly to selected jurisdictions through the Fiscal Year (FY) 2003 Urban Areas Security Initiative (UASI). This financial assistance is being provided to address the unique equipment, training, planning and exercise needs of large urban areas, and to assist them in building an enhanced and sustainable capacity to prevent, respond to, and recover from threats or acts of terrorism. This program also includes a jurisdictional assessment and strategy development component, which will be used by ODP to guide delivery of direct services in the form of training, exercises and technical assistance. This assessment, which must be coordinated with all contiguous jurisdictions and mutual aid partners, will support the development of an Urban Area Homeland Security Strategy, which will in turn provide a roadmap for sustainable prevention and preparedness and guide allocation of grant funds and direct services to the selected jurisdictions. In addition to grant funds and direct services, the cities will be eligible to serve as research and development technology test beds through the Urban Areas Security Initiative. They will also be eligible for technical assistance and support through this program. This program will create a sustainable national model program whereby cities can share the lessons learned and best practices with other cities around the nation.

The Office for Domestic Preparedness transferred from the U.S. Department of Justice, Office of Justice Programs (OJP), to the U.S. Department of Homeland Security effective March 1, 2003. The OJP Office of the Comptroller (OC) will continue to provide fiscal support and oversight to the UASI for the effective period of performance for the grant.

The Fiscal Year 2003 Urban Areas Security Initiative

A.   Authorized Program Purpose

Funding for the FY 2003 UASI is authorized by Public Law 108–7, the Omnibus Appropriations Act of 2003. The FY 2003 UASI seeks to address the unique needs of large urban areas.

ODP also recognizes that the process of a large-scale jurisdictional assessment and development of an Urban Area Homeland Security Strategy is not without cost. Cities may use a portion of the allocated funds to cover expenses associated with the jurisdictional assessment and strategy development, as well as implementation of the UASI. The jurisdictional assessment must be developed in coordination with the State Administrative Agency (SAA) for ODP, who has overall responsibility for developing the State Homeland Security Strategy.

ODP will provide support services to assist you in completing your application. Please consult your respective Program Manager using ODP's toll-free number (1-800-368-6498).

B.   Program Requirements

FY 2003 UASI grantees will be able to initially receive up to 25% of their allocation of funds for the conduct of an assessment, development of a strategy, and management and administration of the program, and to resource immediate needs. Award of these funds is contingent upon: 1) submission and approval of the jurisdiction's online application for funding under the FY 2003 Urban Areas Security Initiative, 2) a written request for funds not to exceed 25% of the total FY 2003 UASI allocation, and 3) submission and approval of an itemized justification for use of those funds. The initial funds are available immediately upon grant award.

Receipt of the remainder of the funds is contingent upon: 1) submission and approval of a jurisdictional assessment, which includes threats, vulnerabilities, needs and capabilities, and has been coordinated with contiguous jurisdictions and mutual aid partners; 2) submission and approval of an Urban Area Homeland Security Strategy articulating a strategic vision for building and sustaining an enhanced level of preparedness and response capacity, and which has been coordinated with the contiguous jurisdictions and mutual aid partners; and 3) submission and approval of budget detail worksheets for all remaining funds. **Applications for funding under this program must be submitted by June 16, 2003, or within 45 days May 1, 2003, the date the solicitation is being posted in OJP's web-based Grants Management System (GMS).**

Jurisdictions may use a recently updated or completed (within the past 12 months) assessment and strategy for the purposes of USAI. Required elements of the jurisdictional assessment and strategy are outlined in Appendix B of this application, which will be used to validate all assessments submitted for review. If a jurisdiction has not completed an assessment, ODP will provide technical assistance to conduct and complete an assessment.

> *ODP is currently working with the states to coordinate a thorough Threats/Vulnerabilities/Needs/Capabilities Assessment and development of a State Homeland Security Strategy for Fiscal Year 2004 funds. The jurisdictional tool developed for this process can also be used by the UASI participants.*
> *In addition, ODP will provide programmatic and budget information in the UASI to State Administrative Agencies (SAAs) in the host states to ensure coordination with the State Homeland Security Strategy.*

Note: Receipt of all remaining funds is contingent upon submission of a validated jurisdictional assessment and a validated Urban Area Homeland Security Strategy.

Each city will be contacted by an ODP program manager to coordinate the submission and/or scheduling of the jurisdictional assessment. If there is an existing assessment, cities will submit that assessment directly to the ODP program manager for review. If the jurisdiction would like technical assistance in developing a jurisdictional assessment or strategy, the ODP program manager will make those arrangements.

Creation of a multi-jurisdictional and multi-disciplinary Urban Area Working Group, with members from all contiguous jurisdictions and mutual aid partners, to develop and implement the program is required. Cities must plan to incorporate contiguous jurisdictions and mutual aid partners into the assessment, strategy development, and any direct services that are delivered by ODP. The leveraging of previously established local working groups is encouraged.

Applications lacking complete information will be accepted and grants awarded, however draw down restrictions will apply until the receipt of all supporting documentation. Restrictions will be rescinded and funds will be released as information is received.

C.   Authorized Program Expenditures

1. **Planning:** Funds to be used for planning may be used to pay for activities associated with the completion of the jurisdictional assessment, development of the Urban Area Homeland Security Strategy, and other planning activities, including: 1) conducting training sessions to ensure accurate completion of the assessment; 2) implementing and managing sustainable programs for equipment acquisition, training and exercises; 3) enhancement or establishment of mutual aid agreements; 4) enhancement or development of emergency operations plans and operating procedures; 5) development or enhancement of recovery plans; 6) development of communication and interoperability protocols and solutions; 7) coordination of citizen and family preparedness plans and programs, including donations programs and volunteer initiatives; 8) enhancement or development of continuity of operations and continuity of government plans; and 9) hiring of full or part-time staff or contractors/consultants to assist with any of the above activities.

2. **Equipment Acquisition:** Funds for equipment purchase may be used to enhance the capabilities of local first responders through the acquisition of: 1) personal protective equipment (PPE); 2) explosive device mitigation and remediation equipment; 3) CBRNE search and rescue equipment; 4) interoperable communications equipment; 5) detection equipment; 6) decontamination equipment; 7) physical security enhancement equipment; 8) terrorism incident prevention equipment; 9) CBRNE logistical support equipment; 10) CBRNE incident response vehicles; 11) medical supplies and limited types of pharmaceuticals; and, 12) CBRNE reference materials. *These funds may also be used for sustainment of first responder equipment that would be used in a jurisdiction's response to a terrorist threat or event. This would include repair and replacement parts, equipment warranties and maintenance contracts for equipment purchased under any ODP grant.*

4

This equipment may be used by the jurisdictions to enhance capabilities in the areas of law enforcement, emergency medical services, emergency management, the fire service, hazardous materials, public works, governmental administrative, public safety communications, healthcare and public health at the local levels of government in accordance with the goals and objectives identified in their jurisdictional assessment. In addition, localities may use their funds to purchase equipment designed to protect critical infrastructure from terrorist attacks and for equipment to prevent terrorist incidents. Sustainment, maintenance, and training in the use of equipment procured using these grant funds should be addressed in the Urban Area Homeland Security Strategy.

3. **Training:** Funds to be used for training may be used to enhance the capabilities of local first responders through the development of a jurisdiction homeland security training program or the delivery of existing ODP courses. Allowable expenses include: 1) development and/or establishment of CBRNE training courses, which should be institutionalized within existing training academies, universities or junior colleges. These courses must be consistent with ODP responder training guidelines and reviewed and approved by ODP; 2) backfill costs to replace responders who are attending ODP-approved courses, 3) overtime costs for responders who attend ODP-approved training courses, and 4) travel costs associated with planning or attending ODP-approved training. Cities should also work with contiguous jurisdictions and mutual aid partners to develop and implement a sustainable system for maintaining and perpetuating training within all response disciplines as part of the overall Urban Area Homeland Security Strategy.

The target audience for training courses funded through the FY 2003 UASI must be emergency responders, emergency managers and public/elected officials within the following disciplines: firefighters, law enforcement, emergency management, emergency medical services, hazardous materials, public works, public health, health care, public safety communications, governmental administrative and private security guards. *We encourage cities to adopt current ODP training programs.*

4. **Exercises:** Funds to be used for exercises may be used to plan for, design, develop, conduct and evaluate exercises that train emergency responders and assess the readiness of jurisdictions to prevent and respond to a terrorist attack. Exercises must be threat- and performance-based, in accordance with ODP's Homeland Security Exercise and Evaluation Program (HSEEP) manuals. These manuals will provide explicit direction on the design, conduct and evaluation of terrorism exercises. Exercises conducted with ODP support (grant funds or direct support) must be managed and executed in accordance with the HSEEP.

ODP will work with cities to build a tailored exercise program. Cities should consider using a portion of the grant funds to convene a multi-disciplinary, multi-jurisdictional exercise planning workshop. Cities should also work with contiguous jurisdictions and mutual aid partners to establish a sustained, multi-jurisdictional cycle of exercise activities as part of the overall Urban Area

Homeland Security Strategy.

4. Allowable exercise-related costs include: 1) expenses related to convening an exercise planning workshop; 2) hiring of full or part-time staff or contractors/consultants to support exercise activities; 3) overtime for first response/exercise management personnel involved in the planning and conducting of exercises; 4) travel associated with planning and conducting exercises; 5) supplies consumed during the course of planning and conducting exercises; 6) costs related to the implementation of the HSEEP to include the reporting of scheduled exercises, and the tracking and reporting of after action reports and corrective actions from exercises; and, 7) other costs related to the planning and conducting of exercise activities.

5. **Management and Administration:** Up to 3% of the grant award may be used to pay for activities associated with the implementation of the overall UASI, including: 1) hiring of full or part-time staff or contractors/consultants to assist with the collection of the assessment data; 2) travel expenses; 3) meeting-related expenses; 4) conducting local program implementation meetings; 5) hiring of full or part-time staff or contractors/ consultants to assist with the implementation and administration of the jurisdictional assessment; and, 6) the acquisition of authorized office equipment. (Note: Authorized office equipment includes personal computers, laptop computers, printers, LCD projectors and other equipment or software which may be required to support implementation of the jurisdiction assessment. For a complete list of allowable meeting-related expenses, please review the OJP Office of the Comptroller (OC) Financial Guide at: http://www.ojp.usdoj.gov/FinGuide.)

D.   Unauthorized Program Expenditures

Unauthorized program expenditures include: 1) expenditures for items such as general-use software (word processing, spreadsheet, graphics, etc), general use computers and related equipment, general-use vehicles, licensing fees, weapons systems and ammunition; 2) activities unrelated to the completion and implementation of the UASI; 3) other items not in accordance with the Authorized Equipment List or previously listed as allowable costs.

III.   Eligible Applicants and Funding Availability

The following jurisdictions are eligible to apply for and receive FY 2003 UASI funding:

CITY NAME	TOTAL
New York City	24,768,000.00
National Capital Region	18,081,000.00
Los Angeles	12,422,000.00
Seattle	11,201,000.00
Chicago	10,896,000.00

| San Francisco | 10,349,000.00 |
| Houston | 8,634,000.00 |

The Chief Executive of each city will designate a program administering agency, also known as the PAA, to apply for and administer the grant funds and administer the Urban Areas Security Initiative. The PAA will coordinate development and implementation of the program with representatives from contiguous and jurisdictions and mutual aid partners through the Urban Area Working Group. The recipient jurisdiction may elect to sub-grant FY 2003 funding directly to other supporting cities or counties with which the jurisdiction has comprehensive mutual aid agreements, or which are part of the overall urban area.

Funding for this program was determined by formula using a combination of current threat estimates, critical assets within the urban area, and population density. The formula is a weighted linear combination of each factor, the result of which is ranked and used to calculate the proportional allocation of resources.

IV.   Application Guidance

Applications need to be prepared according to the directions contained in Section IV and Section V of this booklet.

ODP only accepts applications electronically through the Grant Management System (GMS) located on the Office of Justice Programs (OJP) web site. Instructions regarding electronic submissions through GMS are provided on the OJP web site at www.ojp.usdoj.gov/fundopps.htm. Assistance with GMS may also be obtained by contacting ODP at: 1-800-368-6498.

Questions regarding the application process, equipment procurement issues, issues related to exercises, training, planning and administrative, and programmatic matters involving application submission requirements, application content requirements and other administrative inquiries relating to the FY 2003 UASI can be directed to the ODP Helpline at 1-800-368-6498.

Note: To assist grantees with equipment purchases, ODP has established a number of support programs. These include: 1) the ODP Help Line; 2) the Equipment Purchase Assistance Program; and, 3) the Domestic Preparedness Equipment Technical Assistance Program (DPETAP). The ODP Help Line (1-800-368-6498) provides grantees with answers to non-emergency equipment questions. Likewise, the Equipment Purchase Assistance Program provides ODP grantees with access to GSA prime vendors through memoranda of agreement with the Defense Logistics Agency and the Marine Corps Systems Command. Finally, the DPETAP program provides on-site training in the use and maintenance of specialized response equipment. Additional information on each of these programs can be found on the ODP web site located at: http://www.ojp.usdoj.gov/odp

V.   Application Requirements

A.   On-Line Application: The on-line application must be completed and submitted by the applicant using the OJP GMS system described above. This on-line application replaces the following previously required paper forms:

- Standard Form 424, Application for Federal Assistance
- Standard Form LLL, Disclosure of Lobbying Activities
- OJP Form 4000/3, Assurances
- OJP Form 4061/6, Certifications
- Non-Supplanting Certification

When completing the on-line application, applicants should identify their submissions as new, non-construction applications. These grants are offered by the U.S. Department of Homeland Security. The program title listed in the Catalog of Federal Domestic Assistance (CFDA) is "Office for Domestic Preparedness Fiscal Year 2003 Urban Areas Security Initiative." When referring to this title, please use the following CFDA number: 16.011. *The project period will be for a period not to exceed 24 months.* *(Note: ODP understands the difficulty of procuring equipment from limited supplies and will continue to monitor the industry and make adjustments to project periods as necessary.)*

Note: To expedite the application and award process, no budget information or program narrative is required to apply for this grant. However, applicants MUST provide budget detail worksheets for the initial allocation of funds (up to 25%) accompanying the Categorical Assistance Progress Report due after June 30, 2003. Additional information on this requirement is provided in Appendix D. Applicants will also be required to provide budget detail worksheets for use of the remaining funds after submission and approval of an assessment and urban area strategy, and to provide a final budget prior to closeout of the grant.

B.   Freedom of Information Act (FOIA):   ODP recognizes that much of the information submitted in the course of applying for funding under this program, or provided in the course of its grant management activities, may be considered law enforcement sensitive or otherwise important to national security interests. This may include threat, risk, and needs assessment information, and discussions of demographics, transportation, public works, and industrial and public health infrastructures. While this information is subject to requests made pursuant to the Freedom of Information Act, 5. U.S.C. §552 (FOIA), all determinations concerning the release of information of this nature will be made on a case-by-case basis by the DHS Departmental Disclosure Officer, and may well likely fall within one or more of the available exemptions under the Act.

Applicants are also encouraged to consult their own state and local laws and regulations regarding the release of information, which should be considered when reporting sensitive matters in the grant application, needs assessment and strategic planning process. At the same time, applicants should be aware that any information created exclusively for the purpose of applying for and monitoring grants hereunder is the property of the U.S. Government, and shall not otherwise be disclosed or released pursuant to state or local law or regulation.

VI. Administrative Requirements

A. **Single Point of Contact (SPOC) Review:** Executive Order 12372 requires applicants from state and local units of government or other organizations providing services within a state to submit a copy of the application to the state SPOC, if one exists and if this program has been selected for review by the state. Applicants must contact their state SPOC to determine if the program has been selected for state review. The date that the application was sent to the SPOC or the reason such submission is not required should be provided.

B. **Financial Requirements**

1. Non-Supplanting Certification: This certification, which is a required component of the on-line application, affirms that federal funds will be used to supplement existing funds, and will not replace (supplant) funds that have been appropriated for the same purpose. Potential supplanting will be addressed in the application review as well as in the pre award review, post award monitoring, and the audit. Applicants or grantees may be required to supply documentation certifying that a reduction in non-federal resources occurred for reasons other than the receipt or expected receipt of federal funds.

2. Match Requirement: None

3. Assurances: The on-line application includes a list of assurances that the applicant must comply with in order to receive federal funds under this program. It is the responsibility of the recipient of the federal funds to fully understand and comply with these requirements. Failure to comply may result in the withholding of funds, termination of the award, or other sanctions. You will be agreeing to these assurances when you submit your application on-line through GMS.

4. Certifications Regarding Lobbying; Debarment, Suspension, and Other Responsibility Matters; and Drug-Free Workplace Requirement: This certification, which is a required component of the on-line application, commits the applicant to compliance with the certification requirements under 28 CFR part 69, New Restrictions on Lobbying, and 28 CFR part 67, Government-wide Debarment and Suspension (Nonprocurement) and Government-wide Requirements for Drug-Free Workplace (Grants). The certification will be treated as a material representation of the fact upon which reliance will be placed by the U.S. Department of Homeland Security in awarding grants.

5. Suspension or Termination of Funding: DHS may suspend or terminate funding, in whole or in part, or other measures may be imposed for any of the following reasons:

9

- Failing to comply with the requirements or statutory objectives of federal law.

- Failing to make satisfactory progress toward the goals or objectives set forth in this application.

- Failing to follow grant agreement requirements or special conditions.

- Proposing or implementing substantial plan changes to the extent that, if originally submitted, the application would not have been selected for funding.

- Failing to submit required reports.

- Filing a false certification in this application or other report or document.

Before taking action, ODP will provide the grantee reasonable notice of intent to impose measures and will make efforts to resolve the problem informally.

VII. Reporting Requirements

The following reports are required of all program participants:

- **Financial Status Reports (Standard Form 269A):** Financial Status Reports are due within 45 days of the end of each calendar quarter. A report must be submitted for every quarter the award is active, including partial calendar quarters, as well as for periods where no grant activity occurs (see Appendix L). The OJP OC will provide a copy of this form in the initial award package. Future awards and fund drawdowns will be withheld if these reports are delinquent. The final financial report is due 120 days after the end date of the award period.

- **Program Progress Reports:** Program Progress Reports are due within 30 days after the end of the reporting periods, which are June 30 and December 31, for the life of the award. *Guidance on specific topics to be addressed is included in Appendix D.* The OJP OC will provide a copy of this form in the initial award package. Future awards and fund drawdowns will be withheld if these reports are delinquent. The final programmatic progress report is due 120 days after the end date of the award period.

- **Financial and Compliance Audit Report:** Recipients that expend $300,000 or more of Federal funds during their fiscal year are required to submit an organization-wide financial and compliance audit report. The audit must be performed in accordance with the U.S. General Accounting Office Government Auditing Standards and OMB Circular A-133. Audit reports are currently due to the Federal Audit Clearinghouse no later than

10

9 months after the end of the recipient's fiscal year. In addition, the Secretary of Homeland Security and the Comptroller General of the United States shall have access to any books, documents, and records of recipients of FY 2003 UASI assistance for audit and examination purposes, provided that, in the opinion of the Secretary of Homeland Security or the Comptroller General, these documents are related to the receipt or use of such assistance. The grantee will also give the sponsoring agency or the Comptroller General, through any authorized representative, access to and the right to examine all records, books, papers or documents related to the grant.

Required Performance Related Information: To insure compliance with the Government Performance and Results Act, Public Law 103-62, this solicitation notifies applicants that ODP's performance under this solicitation is measured by: 1) the number of local jurisdictions that enhanced their capacity to respond to a CBRNE incident with new equipment and training; and, 2) the number of jurisdictions that tested preparedness through participation in tabletop or full-scale exercises.

Appendix A

Funding Allocations

URBAN AREAS SECURITY INITIATIVE
FISCAL YEAR 2003 FUNDING ALLOCATIONS

CITY NAME	TOTAL
New York City	24,768,000.00
National Capital Region	18,081,000.00
Los Angeles	12,422,000.00
Seattle	11,201,000.00
Chicago	10,896,000.00
San Francisco	10,349,000.00
Houston	8,634,000.00

Appendix B

Assessment and Strategy Validation

ODP will use the following checklist to review existing jurisdictional assessments for the purposes of this program. The major areas of review are coordination, risk assessment, capabilities assessment, needs assessment, and strategy. *If any or all of these major areas are incomplete or missing, ODP will work with the grantee to craft a valid jurisdictional assessment and strategy.*

COORDINATION:

Jurisdiction has completed major areas one through four in coordination with all contiguous jurisdictions and mutual aid partners through an Urban Area Working Group. Please list the jurisdictions represented on the working group:

_____

_____

_____

MAJOR AREA 1: Risk Assessment

Jurisdiction has completed a risk assessment containing <u>ALL</u> of the following elements:

Threat Assessment - The purpose of a threat assessment is to determine the relative likelihood of a known potential threat element attempting to attack using a weapon of mass destruction.

Identify Potential Threat Elements (PTE) within the Jurisdiction - Potential threat elements are any group or individual in which there are allegations or information indicating a possibility of the unlawful use of force or violence, specifically the utilization of WMD, against persons or property to intimidate or coerce a government, the civilian population, or any segment thereof, in furtherance of a specific motivation or goal, possibly political or social in nature. *Following are examples of threat element factors:*

The Presence of Threat Factors such as:
PTE existence
PTE violent history
PTE intentions
PTE WMD capabilities, and
PTE targeting
Motivation of each PTE
Identification of the WMD capabilities of PTE

Vulnerability Assessment - Basic vulnerability assessments provide local jurisdictions with a current vulnerability profile for all potential targets located within jurisdictional boundaries.

Identify Potential Targets to include: critical infrastructure facilities, sites, systems, or special events that are present or take place within the jurisdiction. *For each of these identified consider:*

Level of Visibility

Criticality of Target Site to Jurisdiction

Impact Outside of Jurisdiction

Potential Threat Element Access to Target

Potential Target Threat of Hazard

Potential Target Site Population Capacity

Potential Collateral Mass Casualties

Rate Target

Identify Legal Hazardous Sites within the following five areas located within Jurisdiction.

Chemical

Biological

Radiological

Nuclear

Explosive

MAJOR AREA 2: Capabilities Assessment

Jurisdiction has completed a capabilities assessment containing <u>ALL</u> of the following elements:

A jurisdictional capabilities assessment examines the current capabilities of the jurisdiction's emergency responders. An effective capabilities assessment will identify the planning, organization, equipment, training, and exercises available to safely and effectively respond to WMD incidents. *The following tasks should be considered in a capabilities assessment:*

Establish CBRNE materials likely to be used during a WMD terrorism incident in order to help determine current capabilities of each response discipline to respond appropriately.

Establish a numerical focus of affected civilians/ responders who may become incapacitated during a WMD terrorism incident.

Apply these factors to planning, organization, equipment, training, and exercises in order to identify gaps.

MAJOR AREA 3: Needs Assessment

Jurisdiction has completed a needs assessment containing <u>ALL</u> of the following elements:

A needs assessment should use the data collected through the risk and capabilities assessments to determine what additional capabilities are needed to increase emergency responder capabilities to respond to a WMD terrorism incident effectively. *The following solution areas should be addressed as part of an adequate needs assessment:*

Planning: Assess planning the jurisdiction has conducted and identify additional requirements to ensure emergency responders have an updated emergency response plan and terrorism incident annex to provide direction in the event of a WMD incident.

Organization: Assess organizational efforts the jurisdiction has addressed through the construction of emergency response teams and written mutual aid agreements with neighboring jurisdictions to provide coverage to those areas lacking WMD response capabilities, and identify additional steps which may be necessary.

Equipment: Assess the equipment resources necessary to respond to WMD terrorism incidents that may occur in the jurisdiction determined through planning factors, WMD response levels by discipline, and specific tasks desired by each discipline.

Training: Assess additional WMD training needed by each discipline using training guidelines provided for increased capabilities.

Exercise: Assess additional exercises the jurisdiction needs to properly exercise their local plans.

MAJOR AREA 4: Strategy Goals, Objectives and Implementation Steps

Jurisdiction has developed goals, objectives and implementation steps containing <u>ALL</u> of the following elements:

The information and data gathered from the jurisdiction identifies current capabilities and projected needs. Using this information, a comprehensive assessment should result in the development of broad-based goals that address areas of response enhancement as well as objectives for each goal. Plans, organizations, equipment, training and exercises are areas for consideration in reducing shortfalls in response capabilities.

A.   Goal: A target that the jurisdiction wants to achieve regarding an improved level of capability.
> Help jurisdiction to achieve its vision
> Focuses on the long term
> Broad in scope

B.   Objective: Specific statements of desired achievement that supports the goal.
> Support the attainment of long range goals
> Focus on short term
> It is specific, measurable, achievable, results-oriented, and time-limited

C.   Implementation step: A road map to the accomplishment of the objective.

D.   Metric(s): A tool for measuring progress in achieving goals and objectives.

Appendix C

Allowable Planning, Equipment, Training, Exercise, and Administrative Costs

I. Allowable Planning Costs

Funds to be used for planning may be used to pay for activities associated with the completion of the jurisdictional assessment, development of the Urban Area Homeland Security Strategy, and other planning activities, including: 1) conducting training sessions to ensure accurate completion of the assessment; 2) implementing and managing sustainable programs for equipment acquisition, training and exercises; 3) enhancement or establishment of mutual aid agreements; 4) enhancement or development of emergency operations plans and operating procedures; 5) development or enhancement of recovery plans; 6)development of communication and interoperability protocols and solutions; 7) coordination of citizen and family preparedness plans and programs, including donations programs and volunteer initiatives; 8) enhancement or development of continuity of operations and continuity of government plans; and 9) hiring of full or part-time staff or contractors/consultants to assist with any of the above activities.

II. Allowable Equipment Costs

Funds from the Urban Areas Security Initiative may be used by the cities to enhance capabilities in the areas of law enforcement, emergency medical services, emergency management, the fire service, hazardous materials, public works, governmental administrative, public safety communications, healthcare and public health at the state and local levels of government in accordance with the goals and objectives identified in jurisdictional assessment or capability enhancement plan. In addition, cities may use their funds to purchase equipment designed to protect critical infrastructure from terrorist attacks and for equipment to prevent terrorist incidents. Funds may also be used for sustainment of first responder equipment that would be used in a jurisdiction's response to a terrorist threat or event. This would include repair and replacement parts, equipment warranties and maintenance contracts for equipment purchased under this ODP grant.

Authorized equipment purchases may be made in the following categories:

1. Personal Protective Equipment (PPE)
2. Explosive Device Mitigation and Remediation Equipment
3. CBRNE Search & Rescue Equipment
4. Interoperable Communications Equipment
5. Detection Equipment
6. Decontamination Equipment
7. Physical Security Enhancement Equipment
8. Terrorism Incident Prevention Equipment
9. CBRNE Logistical Support Equipment
10. CBRNE Incident Response Vehicles
11. Medical Supplies and Limited Types of Pharmaceuticals
12. CBRNE Reference Materials

1. Personal Protective Equipment -   Equipment worn to protect the individual from hazardous materials and contamination. Levels of protection vary and are divided into three categories based on the degree of protection afforded. The following constitutes equipment intended for use in a chemical/biological threat environment:

Level A. Fully encapsulated, liquid and vapor protective ensemble selected when the highest level of skin, respiratory and eye protection is required. The following constitutes Level A equipment for consideration:

· Fully Encapsulated Liquid and Vapor Protection Ensemble, reusable or disposable (tested and certified against CB threats)
  Fully Encapsulated Training Suits
· Closed-Circuit Rebreather (minimum 2-hour supply, preferred), or open-circuit Self-Contained Breathing Apparatus (SCBA) or, when appropriate, Air-Line System with 15-minute minimum escape SCBA
· Spare Cylinders/Bottles for rebreathers or SCBA and service/repair kits
· Chemical Resistant Gloves, including thermal, as appropriate to hazard

  Personal Cooling System; Vest or Full Suit with support equipment needed for maintaining body core temperature within acceptable limits
  Hardhat/helmet
· Chemical/Biological Protective Undergarment
· Inner Gloves
· Approved Chemical Resistant Tape
· Chemical Resistant Boots, Steel or Fiberglass Toe and Shank
· Chemical Resistant Outer Booties

Level B. Liquid splash resistant ensemble used with highest level of respiratory protection. The following constitute Level B equipment and should be considered for use:

· Liquid Splash Resistant Chemical Clothing, encapsulated or non-encapsulated
· Liquid Splash Resistant Hood
· Closed-Circuit Rebreather (minimum 2-hour supply, preferred), open-circuit SCBA, or when appropriate, Air-Line System with 15-minute minimum escape SCBA
· Spare Cylinders/Bottles for rebreathers or SCBA and service/repair kits
· Chemical Resistant Gloves, including thermal, as appropriate to hazard
· Personal Cooling System; Vest or Full Suit with support equipment needed for maintaining body core temperature within acceptable limits
· Hardhat/helmet
· Chemical/Biological Protective Undergarment
· Inner Gloves
· Approved Chemical Resistant Tape
· Chemical Resistant Boots, Steel or Fiberglass Toe and Shank
· Chemical Resistant Outer Booties

Level C. Liquid splash resistant ensemble, with same level of skin protection of Level B, used when the concentration(s) and type(s) of airborne substances(s) are known and the criteria for using air-purifying respirators are met. The following constitute Level C equipment and should be considered for use:

- Liquid Chemical Splash Resistant Clothing (permeable or non-permeable)
- Liquid Chemical Splash Resistant Hood (permeable or non-permeable)
- Tight-fitting, Full Facepiece, Negative Pressure Air Purifying Respirator with the appropriate cartridge(s) or canister(s) and P100 filter(s) for protection against toxic industrial chemicals, particulates, and military specific agents.
- Tight-fitting, Full Facepiece, Powered Air Purifying Respirator (PAPR) with chemically resistant hood with appropriate cartridge(s) or canister(s) and high-efficiency filter(s) for protection against toxic industrial chemicals, particulates, and military specific agents.
- Equipment or system batteries will include those that are rechargeable (e.g. NiCad) or non-rechargeable with extended shelf life (e.g. Lithium)
- Chemical Resistant Gloves, including thermal, as appropriate to hazard
- Personal Cooling System; Vest or Full Suit with support equipment
- Hardhat
- Inner Chemical/Biological Resistant Garment
  Inner Gloves
- Chemical Resistant Tape
- Chemical Resistant Boots, Steel or Fiberglass Toe and Shank
- Chemical Resistant Outer Booties

Level D. Selected when no respiratory protection and minimal skin protection is required, and the atmosphere contains no known hazard and work functions preclude splashes, immersion, or the potential for unexpected inhalation of, or contact with, hazardous levels of any chemicals.

- Escape mask for self-rescue

Note: During CBRNE response operations, the incident commander determines the appropriate level of personal protective equipment. As a guide, Levels A, B, and C are applicable for chemical/ biological/ radiological contaminated environments. Personnel entering protective postures must undergo medical monitoring prior to and after entry.

All SCBAs should meet standards established by the National Institute for Occupational Safety and Health (NIOSH) for occupational use by emergency responders when exposed to Chemical, Biological, Radiological and Nuclear (CBRN) agents in accordance with Special Tests under NIOSH 42 CFR 84.63(c), procedure number RCT-CBRN-STP-0002, dated December 14, 2001.

Grant recipients should purchase: 1) protective ensembles for chemical and biological terrorism incidents that are certified as compliant with Class 1, Class 2, or Class 3 requirements of National Fire Protection Association (NFPA) 1994, Protective Ensembles for Chemical/Biological Terrorism Incidents; 2) protective ensembles for hazardous materials emergencies that are certified as compliant with NFPA 1991, Standard on Vapor Protective Ensembles for Hazardous Materials Emergencies, including the chemical and biological terrorism protection; 3) protective ensembles for search and rescue or search and recovery operations where there is no exposure to chemical or biological warfare or terrorism agents and where exposure to flame and heat is unlikely or nonexistent that are certified as compliant with NFPA 1951, Standard on Protective Ensemble for USAR Operations; and, 4) protective clothing from blood and body fluid pathogens for persons providing treatment to victims after decontamination that are certified as compliant with NFPA 1999, Standard on Protective Clothing for Emergency Medical Operations. For more information regarding these standards, please refer to the following web sites:

The National Fire Protection Association - http://www.nfpa.org

National Institute for Occupational Safety and Health - http://www.cdc.gov/niosh

2. Explosive Device Mitigation and Remediation - Equipment providing for the mitigation and remediation of explosive devices in a CBRNE environment:

- Bomb Search Protective Ensemble for Chemical/Biological Response
- Chemical/Biological Undergarment for Bomb Search Protective Ensemble
- Cooling Garments to manage heat stress
- Ballistic Threat Body Armor (not for riot suppression)
- Ballistic Threat Helmet (not for riot suppression)
- Blast and Ballistic Threat Eye Protection (not for riot suppression)
- Blast and Overpressure Threat Ear Protection (not for riot suppression)
- Fire Resistant Gloves
- Dearmer/Disrupter
- Real Time X-Ray Unit; Portable X-Ray Unit
- CBRNE Compatible Total Containment Vessel (TCV)
- CBRNE Upgrades for Existing TCV
- Robot; Robot Upgrades
- Fiber Optic Kit (inspection or viewing)
- Tents, standard or air inflatable for chem/bio protection
- Inspection mirrors
- Ion Track Explosive Detector

3. CBRNE Search and Rescue Equipment - Equipment providing a technical search and rescue capability for a CBRNE environment:

- Hydraulic tools; hydraulic power unit
- Listening devices; hearing protection
- Search cameras (including thermal and infrared imaging)
- Breaking devices (including spreaders, saws and hammers)
- Lifting devices (including air bag systems, hydraulic rams, jacks, ropes and block and tackle)
- Blocking and bracing materials
- Evacuation chairs (for evacuation of disabled personnel)
- Ventilation fans

4. Interoperable Communications Equipment - Equipment and systems providing connectivity and electrical interoperability between local and interagency organizations to coordinate CBRNE response operations:

- Land Mobile, Two-Way In-Suit Communications (secure, hands-free, fully duplex, optional), including air-to-ground capability (as required)
- Antenna systems
- Personnel Alert Safety System (PASS) - (location and physiological monitoring systems optional)
- Personnel Accountability Systems
- Individual/portable radios, software radios, portable repeaters, radio interconnect systems, satellite phones, batteries, chargers and battery conditioning systems
- Computer systems designated for use in an integrated system to assist with detection and communication efforts (must be linked with integrated software packages designed specifically for chemical and/or biological agent detection and communication purposes)
  Portable Meteorological Station (monitors temperature, wind speed, wind direction and barometric pressure at a minimum)
- Computer aided dispatch system
- Commercially available crisis management software
- Mobile Display Terminals

Note: In an effort to improve public safety interoperability, all new or upgraded radio systems and new radio equipment should be compatible with a suite of standards called ANSI/TIA/EIA-102 Phase I (Project 25). These standards have been developed to allow for backward compatibility with existing digital and analog systems and provide for interoperability in future systems. The FCC has chosen the Project 25
suite of standards for voice and low-moderate speed data interoperability in the new nationwide 700 MHZ frequency band and the Integrated Wireless Network (IWN) of the U.S. Justice and Treasury Departments has chosen the Project 25 suite of standards for their new radio equipment. ***In an effort to realize improved interoperability, all radios purchased under this grant should be APCO 25 compliant.***

5. Detection Equipment - Equipment to sample, detect, identify, quantify, and monitor for chemical, biological, radiological/nuclear and explosive agents throughout designated areas or at specific points:

Chemical

- M-8 Detection Paper for chemical agent identification
- M-9 Detection Paper (roll) for chemical agent (military grade) detection
- M-256 Detection Kit for Chemical Agent (weapons grade—blister: CX/HD/L; blood: AC/CK; and nerve: GB/VX) detection
- M-256 Training Kit
- M-18 Series Chemical Agent Detector Kit for surface/vapor chemical agent analysis
- Hazard Categorizing (HAZCAT) Kits
- Photo-Ionization Detector (PID)
- Flame Ionization Detector (FID)
- Surface Acoustic Wave Detector
- Gas Chromatograph/Mass Spectrometer (GC/MS)

- Ion Mobility Spectrometry
- Stand-Off Chemical Detector
- M-272 Chemical Agent Water Test Kit
- Colormetric Tube/Chip Kit specific for TICs and CBRNE applications
- Multi-gas Meter with minimum of O2 and LEL
- Leak Detectors (soap solution, ammonium hydroxide, etc)
- pH Paper/pH Meter
- Waste Water Classifier Kit
- Oxidizing Paper
- Protective cases for sensitive detection equipment storage & transport

Biological

Point Detection Systems/Kits (Immunoassay or other technology)

Radiological/Nuclear

- Radiation detection equipment (electronic or other technology that detects alpha, beta, gamma, and high intensity gamma)
- Personal Dosimeter
- Scintillation Fluid (radiological) pre-packaged
- Radiation monitors

Explosive

- Canines (initial acquisition, initial operational capability only)

6. Decontamination Equipment - Equipment and material used to clean, remediate, remove or mitigate chemical and biological contamination:

Chemical

- Decontamination system for individual and mass application with environmental controls, water heating system, showers, lighting, and transportation (trailer)
- Decon Litters/roller systems
- Extraction Litters, rollable
- Runoff Containment Bladder(s), decontamination shower waste collection with intrinsically-safe evacuation pumps, hoses, connectors, scrub brushes, nozzles
- Spill Containment Devices
- Overpak Drums
- Non-Transparent Cadaver Bags (CDC standard)
- Hand Carts
- Waste water classification kits/strips

Biological

- HEPA (High Efficiency Particulate Air) Vacuum for dry decontamination

7. Physical Security Enhancement Equipment - Equipment to enhance the physical security of critical infrastructure.

Surveillance, Warning, Access/Intrusion Control

Ground

- Motion Detector Systems: Acoustic; Infrared; Seismic; Magnetometers
- Barriers: Fences; Jersey Walls
- Impact Resistant Doors and Gates
- Portal Systems; locking devices for access control
- Alarm Systems
- Video Assessment/Cameras: Standard, Low Light, IR, Automated Detection
  Personnel Identification: Visual; Electronic; Acoustic; Laser; Scanners; Cyphers/Codes
- X-Ray Units
- Magnetometers
- Vehicle Identification: Visual; Electronic; Acoustic; Laser; Radar

Waterfront
- Radar Systems
- Video Assessment System/Cameras: Standard, Low Light, IR, Automated Detection
- Diver/Swimmer Detection Systems; Sonar
- Impact Resistant Doors and Gates
- Portal Systems
- Hull Scanning Equipment
- Plus all those for Ground

Sensors – Agent/Explosives Detection

- Chemical: Active/Passive; Mobile/Fixed; Handheld
- Biological: Active/Passive; Mobile/Fixed; Handheld
- Radiological
- Nuclear
- Ground/Wall Penetrating Radar

Inspection/Detection Systems

- Vehicle & Cargo Inspection System – Gamma-ray
- Mobile Search & Inspection System – X-ray
- Non-Invasive Radiological/Chem/Bio/Explosives System – Pulsed Neutron Activation

Explosion Protection

- Blast/Shock/Impact Resistant Systems
- Protective Clothing

- Column and Surface Wraps; Breakage/Shatter Resistant Glass; Window Wraps
- Robotic Disarm/Disable Systems

8. Terrorism Incident Prevention Equipment (Terrorism Early Warning, Prevention, and Deterrence Equipment and Technologies) - Local public safety agencies will increasingly rely on the integration of emerging technologies and equipment to improve jurisdictional capabilities to deter and prevent terrorist incidents. This includes, but is not limited to, equipment and associated components that enhance a jurisdiction's ability to disseminate advanced warning information to prevent a terrorist incident or disrupt a terrorist's ability to carry out the event, including information sharing, threat recognition, and public/private sector collaboration.

- Data collection/information gathering software
  Data synthesis software
- Geographic Information System information technology and software
- Law enforcement surveillance equipment

9. CBRNE Logistical Support Equipment - Logistical support gear used to store and transport the equipment to the CBRNE incident site and handle it once onsite. This category also includes small support equipment including intrinsically safe (non-sparking) hand tools required to support a variety of tasks and to maintain equipment purchased under the grant as well as general support equipment intended to support the CBRNE incident response:

- Equipment trailers
- Weather-tight containers for equipment storage
- Software for equipment tracking and inventory
- Handheld computers for Emergency Response applications
- Small Hand tools
- Binoculars, head lamps, range finders and spotting scopes (not for weapons use)
- Small Generators to operate light sets, water pumps for decontamination sets
- Light sets for nighttime operations/security
- Electrical Current detectors
- Equipment harnesses, belts, and vests
- Isolation containers for suspected chemical/biological samples
- Bull horns
- Water pumps for decontamination systems
- Bar code scanner/reader for equipment inventory control
- Badging system equipment and supplies
- Cascade system for refilling SCBA oxygen bottles
- SCBA fit test equipment and software to conduct flow testing
- Testing Equipment for fully encapsulated suits
- Cooling/Heating/Ventilation Fans (personnel and decontamination tent use)
- HAZMAT Gear Bag/Box

10.   CBRNE Incident Response Vehicles - This category includes special-purpose vehicles for the transport of CBRNE response equipment and personnel to the incident site. Licensing and registration fees are the responsibility of the jurisdiction and are not allowable under this grant. In addition, general purpose vehicles (squad cars, executive transportation, etc.), fire apparatus, and tactical/armored assault vehicles are not allowable. Allowable vehicles include:

- Mobile command post vehicles
- Hazardous materials (HazMat) response vehicles
- Bomb response vehicles
- Prime movers for equipment trailers
- 2-wheel personal transport vehicles for transporting fully suited bomb technicians, Level A/B suited technicians to the Hot Zone
- Multi-wheeled all terrain vehicles for transporting personnel and equipment to and from the Hot Zone

11. Medical Supplies and Pharmaceuticals - Medical supplies and pharmaceuticals required for response to a CBRNE incident. Grantees are responsible for replenishing items after shelf-life expiration date(s).

Medical Supplies

- Automatic Biphasic External Defibrillators and carry bags
- Equipment and supplies for establishing and maintaining a patient airway at the advanced life support level (to include OP and NG airways; ET tubes, styletes, blades, and handles; portable suction devices and catheters; and stethoscopes for monitoring breath sounds)
- Blood Pressure Cuffs
- IV Administration Sets (Macro and Micro) and Pressure Infusing Bags
- IV Catheters (14, 16, 18, 20, and 22 gauge)
- IV Catheters (Butterfly 22, 24 and 26 gauge)
- Manual Biphasic Defibrillators (defibrillator, pacemaker, 12 lead) and carry bags
- Eye Lense for Lavage or Continuous Medication
- Morgan Eye Shields
- Nasogastric Tubes
- Oxygen administration equipment and supplies (including bag valve masks; rebreather and non-rebreather masks, and nasal cannulas; oxygen cylinders, regulators, tubing, and manifold distribution systems; and pulse oximetry, Capnography & CO_2 detection devices)
- Portable Ventilator
- Pulmonary Fit Tester
- Syringes (3cc and 10cc)
- 26 ga ½" needles (for syringes)
- 21 ga. 1 ½ " needles (for syringes)
- Triage Tags and Tarps
- Sterile and Non-Sterile dressings, all forms and sizes
- Gauze, all sizes

Pharmaceuticals

- 2Pam Chloride

- Adenosine
- Albuterol Sulfate .083%
- Albuterol MDI
- Atropine 0.1 & 0.4 mg/ml
- Atropine Auto Injectors
- Benadryl
- CANA Auto Injectors
- Calcium Chloride
- Calcium Gluconate 10%
- Ciprofloxin PO
- Cyanide kits

- Dextrose
- Dopamine
- Doxycycline PO
- Epinephrine
- Glucagon
- Lasix
- Lidocaine
- Loperamide
- Magnesium Sulfate
- Methylprednisolone
- Narcan
- Nubain
- Nitroglycerin
- Normal Saline (500 and 1000 ml bags)
- Potassium Iodide
- Silver Sulfadiazine
- Sodium Bicarbonate
- Sterile Water
- Tetracaine
- Thiamine
- Valium

12. CBRNE Reference Materials - Reference materials designed to assist emergency first responders in preparing for and responding to a CBRNE incident. This includes but is not limited to the following:

- NFPA Guide to hazardous materials
- NIOSH Hazardous Materials Pocket Guide
- North American Emergency Response Guide
- Jane's Chem-Bio Handbook
  First Responder Job Aids

III. Allowable Training Costs

Funds from the Urban Areas Security Initiative may be used to enhance the capabilities of local emergency responders through the enhancement or development of an urban area homeland security training program, or delivery of existing ODP courses. Allowable training-related costs include: 1) development and/or establishment of CBRNE training courses, which should be institutionalized within existing training academies, universities or junior colleges. These courses must be consistent with ODP responder training guidelines and reviewed and approved by ODP; 2) backfill costs to replace responders who are attending ODP-approved courses; 3) overtime costs for responders who attend ODP-approved training courses, and 4) travel costs associated with planning or attending ODP-approved training.

The target audience for training supported through the FY 2003 UASI must be emergency responders, emergency managers and public/elected officials within the following disciplines: firefighters, law enforcement, emergency management, emergency medical services, hazardous materials, public works, public health, health care, public safety communications, governmental administrative and private security guards. Grantees using these funds to develop their own courses should address the critical training areas and gaps identified in the jurisdictional assessment or capability enhancement plan and must adhere to the ODP Emergency Responder Guidelines. These guidelines may be found at: http://www.ojp.usdoj.gov/odp/whatsnew/whats_new.htm

IV. Allowable Exercise Costs

Funds from the Urban Areas Security Initiative may be used to plan for, design, develop, conduct and evaluate exercises that train emergency responders and assess the readiness of jurisdictions to prevent and respond to a terrorist attack. Exercises must be threat and performance-based, in accordance with ODP's Homeland Security Exercise and Evaluation Program (HSEEP) manuals. These manuals will provide explicit direction on the design, conduct and evaluation of terrorism exercises. Exercises conducted with ODP support (grant funds or direct support) must be managed and executed in accordance with the HSEEP.

Allowable exercise-related costs include:

1. Exercise Planning Workshop - Grant funds may be used to plan and conduct an Exercise Planning Workshop to include costs related to planning, meeting space and other meeting costs, facilitation costs, materials and supplies, travel and exercise plan development.

2. Full or Part-Time Staff or Contractors/Consultants - Full or part-time staff may be hired to support exercise-related activities. Payment of salaries and fringe benefits must be in accordance with the policies of the unit(s) of local government and have the approval of the awarding agency. The services of contractors/consultants may also be procured by the city in the design, development, conduct and evaluation of CBRNE exercises. The applicant's formal written procurement policy or the Federal Acquisition Regulations (FAR) must be followed.

3. Overtime - Payment of overtime expenses will be for work performed by award (PAA) or sub-award employees in excess of the established work week (usually 40 hours). Further, overtime payments are allowed only to the extent the payment for such services is in accordance with the policies of the unit(s) of local government and has the approval of the awarding agency. In no case is dual compensation allowable. That is, an employee of a unit of government may not receive compensation from their unit or agency of government AND from an award for a single period of time (e.g., 1:00 pm to 5:00 pm), even though such work may benefit both activities. Fringe benefits on overtime hours are limited to FICA, Workman's Compensation and Unemployment Compensation.

4. Travel - Travel costs (i.e., airfare, mileage, per diem, hotel, etc.) are allowable as expenses by employees who are on travel status for official business related to the planning and conduct of the exercise project(s). These costs must be in accordance with either the federal or an organizationally-approved travel policy.

5. Supplies - Supplies are items that are expended or consumed during the course of the planning and conduct of the exercise project(s) (e.g., copying paper, gloves, tape, and non-sterile masks).

6. Implementation of the HSEEP - Costs related to setting up and maintaining a system to track the completion and submission of After Action Reports (AAR) and the implementation of Corrective Action Plans (CAP) from exercises, which may include costs associated with meeting with local jurisdictions to define procedures. *(Note: ODP is developing a national information system for the scheduling of exercises and the tracking of AAR/CAPs to reduce the burden on the localities and to facilitate national assessments of preparedness.)*

7. Other Items - These costs include the rental of space/locations for exercise planning and conduct, exercise signs, badges, etc.

V. Management and Administrative Costs

Up to 3% of the grant award may be used to pay for activities associated with the implementation of the overall UASI, including: 1) hiring of full or part-time staff or contractors/consultants to assist with the collection of the assessment data; 2) travel expenses; 3) meeting-related expenses; 4) conducting local program implementation meetings; 5) hiring of full or part-time staff or contractors/ consultants to assist with the implementation and administration of the jurisdictional assessment; and, 6) the acquisition of authorized office equipment. (Note: Authorized office equipment includes personal computers, laptop computers, printers, LCD projectors and other equipment or software which may be required to support implementation of the jurisdiction assessment. For a complete list of allowable meeting-related expenses, please review the OJP Office of the Comptroller (OC) Financial Guide at: http://www.ojp.usdoj.gov/FinGuide.)

Appendix D

Grant Reporting Requirements

Grant Reporting Requirements

I. Financial Status Reports (SF-269A)

A. Reporting Time Line - Financial Status Reports are due within **45** days after the end of each calendar quarter. A report must be submitted for every quarter that the award is active, including partial calendar quarters, as well as for periods where no grant activity occurs (see below).

	Report Period	Report Due By	Report Period	Report Due By	Report Period	Report Due By	Report Period	Report Due By
First Quarter	1/1 - 3/31	5/15						
Second Quarter			4/1 - 6/30	8/14				
Third Quarter					7/1 - 9/30	11/14		
Fourth Quarter							10/1 - 12/31	2/14

II. Categorical Assistance Progress Reports (OJP Form 4587/1)

A. Reporting Time Line - Categorical Assistance Progress Reports are due within **30** days after the end of the reporting periods, which are June and December 31, for the life of the award.

	Report Period	Report Due By	Report Period	Report Due By
First Quarter	1/1 - 6/30	7/31		
Second Quarter				
Third Quarter			7/1 - 12/31	1/31
Fourth Quarter				

B. **Budget Detail Worksheets** - Grantees MUST provide detailed budget information for the initial allocation of funds (up to 25%) provided through the Fiscal Year 2003 Urban Areas Security Initiative. *This information MUST be provided as an attachment to the Categorical Assistance Progress Report due after June 30, 2003.* Additionally, Budget Detail Worksheets for use of the remaining funds under the grant award must be submitted and approved by ODP prior to obligation, expenditure, or draw-down of those funds. Sample Budget Detail Worksheets detailing the information that MUST be furnished for each allocation are provided below:

Sample Budget Detail Worksheet for Direct Purchases of Equipment

Jurisdiction	Category	Item	Quantity	Total Cost	Discipline Allocation**
City Name	PPE	SCBA - 30 min.	10	$500	HZ - $250 LE - $250
City Name	Detection	Chemical Agent Monitor	2	$14,000	HZ- $7,000 FS - $7000
City Name	Communications	Radio Interconnect System	3	$150,000	FS - $50,000 LE - $50,000 EMA - $50,000
				Total $164,500	

** Law Enforcement (LE), Emergency Medical Services (EMS), Emergency Management (EMA) Fire Service (FS), HazMat (HZ), Public Works (PW), Public Health (PH), Governmental Administrative (GA), Public Safety Communications (PSC), Health Care (HC).

Sample Budget Detail Worksheet for Direct Purchases of Exercise, Training, Planning, and Administrative Services

Jurisdiction	Function	Category	Item	Amount
City Name	Training	Overtime	Support attendance at Incident Response to Terrorist Bombings Course	$750
City Name	Exercises	Contractor	Design CBRNE Exercise Program	$130,000
City Name	Planning	Personnel	Grant Manager	$50,000
City Name	Admin	Travel	Conference Expense	$200
			Total	$135,950

Sample Budget Detail Worksheet for Sub-Awards

Jurisdiction	Equipment Allocation	Exercise Allocation	Training Allocation	Planning/ Admin Allocation
City Name	$200,000.00		$150,000.00	
City Name	$800,000.00	$150,000.00		
City Name		$300,000.00	$300,000.00	$250,000.00
Total	$1,000,000.00	$450,000.00	$450,000.00	$250,000.00

C. Additional Information - Grantees must also use Block 12 of *each* Categorical Assistance Progress Report to describe progress to date in implementing the grant and its impact on homeland security in the state. Each report must provide an update on the following activities that occurred during the designated reporting period:

1) Describe progress made to date in implementing this grant for each of the areas (planning, equipment, training, exercises, and administration).

Planning:
Report hiring of additional staff and activities they have pursued. Additionally, indicate steps taken to facilitate the jurisdictional assessment, development of the Urban Area Homeland Security Strategy, and other planning activities undertaken.

Equipment
If your city is sub-granting funds, report:

- The total number of sub-grants that the city intends to award;
- The number of sub-grants that the city has awarded to date, with the total amount of awards made thus far;
- The names of agencies and/or jurisdictions that have received sub-grant awards during the reporting period and the amounts received; and
- The total amount of funds expended through approved sub-grants to date.
- Attach any new equipment budgets that have been submitted to the city by sub-grantees during the reporting period. Each sub-grant budget should include the Equipment Budget Category, Item, Quantity, Estimated Total Cost and Discipline(s) receiving the equipment (see Appendix C).

If your city is purchasing and distributing equipment, report:

- The city agencies and/or local jurisdictions that have received equipment during the reporting period and the funding amount allocated for each;
- The total amount of funding that has been obligated by the city thus far;

- The total amount of funding that has been expended by the city thus far; and,
- The percentage of overall equipment (in dollar value) that has been received and distributed.

Regardless of the city's method of funding/equipment distribution, indicate whether personnel within recipient agencies and/or jurisdictions are sufficiently trained to use grant-funded equipment, and if technical assistance or other training is needed. Any procurement, distribution, or other equipment related problems should also be noted in the progress report.

Training:
Provide information on how training funds have been used. Some relevant questions to ask include:

- Are funds being used to develop a comprehensive Homeland Security training program in the city?
- What steps have been identified and taken?
- Has training staff been hired? What activities are they undertaking?
- Have funds been provided to state academies, universities, or other institutions to enhance Homeland Security preparedness? How will this impact the number of responders and other officials trained?
- If funds have been provided to academies, universities, or other institutions, how many individuals have been trained?
- Are funds being provided to local jurisdictions and city agencies to pay for training courses/overtime?

Exercises
If the city is using exercise funds to hire city-level staff/contractors, indicate general activities that have occurred during the reporting period. Describe how the additional staff has contributed to enhancement of exercise programs within the city. Relevant questions to address may include:

- What elements of a citywide exercise program have been developed?
- What type of coordination has occurred between city-level exercise staff/contractors and local jurisdictions and/or mutual aid partners?
- What exercises have city exercise staff/contractors helped identify, develop, conduct, and/or evaluate? What jurisdictions were involved?
- If exercises have occurred, how are the post-evaluations being used to identify and address preparedness needs?
- Have actions been taken on any exercise evaluation findings?
- Attach any outstanding after action reports.

If exercise funds were provided to local jurisdictions and/or mutual/aid partners to develop, conduct, assess, and/or participate in exercises, indicate how those funds were generally used. Relevant questions to address may include:

- What jurisdictions/agencies were involved in the exercises?

- What activities did the funds support (i.e. overtime for participants, contractors, etc.)?
- If exercises have occurred, how are the post-evaluations being used to identify and address preparedness needs?
- Have actions been taken on any exercise evaluation findings?

Administration:
Report hiring of additional staff and activities they have pursued. Describe meetings and assessment training sessions that have occurred with city officials and contiguous jurisdictions and mutual aid partners. Note any difficulties and indicate if technical assistance is needed.

2) Additionally, use the Progress Report to:

- Describe progress made to date on achieving the city's overall goals and objectives as identified in the Urban Area Homeland Security Strategy.

- Briefly explain how ODP resources (other than those already addressed above) are contributing to attaining the overall goals and objectives identified in the jurisdictional assessment.

- Identify other significant activities/ initiatives your city and/or local jurisdictions are pursuing to enhance overall preparedness and responder capabilities, particularly those initiatives not previously addressed in the jurisdictional assessment, or not being supported by ODP resources.

- Identify problems your city is encountering regarding the implementation of any area of the jurisdictional assessment or capability enhancement plan, and any steps taken by your city to resolve these problems/ issues.

- If applicable, briefly describe any unique initiatives/ promising practices your city has undertaken that may be applicable to other cities or jurisdictions.

- If applicable, provide any feedback on the ODP grant process and ODP assistance with program implementation, including implementation of the jurisdictional assessment. Identify any other issues or concerns not addressed above.

Chapter 7 Appendix

AMERICA'S SECURITY SINCE 9/11

FACT SHEET: AN OVERVIEW OF AMERICA'S SECURITY SINCE 9/11

The country has made great strides toward improving the security of our homeland since September 11th. Whether by land, sea, or air, it is now substantially more difficult for terrorists to enter the United States; homeland security professionals are sharing information like never before; and America's citizens are better prepared for a natural disaster or terrorist attack. A snapshot of the ways we are safer today:

Curb to Cockpit: Air travel is safer now than ever before due to the layered security DHS has put in place hardened cockpit doors on 100% of large passenger aircraft, vulnerability assessments at over 75 of the nation's largest airports, 100% of all checked baggage is screened, deployment of thousands of federal air marshals and a professionally trained screener workforce which has intercepted more than 12.4 million prohibited items since their inception. In addition, a robust screening system is in place for all international flights into the United States, and all passenger names for domestic flights are checked against an expanded terrorist watch lists.

Port to Port: New security measures specifically tailored to the individual port are now in place at every port in America. These layered measures begin overseas by screening cargo before it's loaded on ships in foreign ports. Homeland Security screens 100% of high risk cargo by targeting suspect cargo using a set of specific indicators. Every port in America has submitted a security plan which includes security measures such as surveillance cameras, background checks on port workers. The Department is also moving forward to implement "smart" technologies for cargo containers. Secure Borders and Open Doors: The Department of Homeland Security has launched the US-VISIT system which links databases to provide valuable information to port of entry officials and consular officials overseas and creates a database of pictures and finger scans of everyone entering the United States with a non-immigrant visa (and soon to include visa waiver travelers). This new tool means that we have a much better idea of who is entering our country. If a traveler's finger scan hits a match on the terrorist watch list, the Department is able to stop them from entering the country at the border. Over 200 people have already been turned away from our borders using this new system. Increased Information Sharing: Several information sharing vehicles exist today that did not exist before September 11, 2001. The Homeland Security Information Network, which is available in all 50 states, makes threat-related information available to law enforcement and emergency managers on a daily basis through a web-based system. Members of the private sector now receive threat-related information through the HSIN system. In addition, members of 35 different Federal agencies are now all co-located together in DHS's new 24-hour Homeland Security Operations Center, which allows the information coming from various sources to be synthesized together and then shared with other federal partners such as the FBI and the Department of Defense. In addition, nearly 100 bulletins and other threat related communiqués have been sent to homeland security professionals across the country.

Citizen Preparedness: September is National Preparedness Month. More than 80 partners and all 56 states and territories are making individual and family preparedness a priority across the nation by hosting events, offering training sessions and distributing information. In addition, the public education campaign Ready and its Spanish language version Listo educates and empowers American citizens to prepare for and respond to potential terrorist attacks and other emergencies. Ready, the most successful public service campaign launched in Ad Council history, delivers its messages through the www.Ready.gov and www.Listo.gov websites, radio, television, print and outdoor PSAs, brochures and a variety of partnerships with private sector organizations. Ready Business will be launched later this month to encourage small- to medium-sized businesses to take steps to safeguard their employees and assets while preparing for business continuity in the event of a disaster. Also, more than 1,300 communities around the country, encompassing 50 percent of the U.S. population, have established Citizen Corps Councils to engage citizens in preparing, training and volunteer service, including delivering the important messages of the Ready campaign. Interoperability: DHS's Safecom program provides long-term technical assistance to federal, state, tribal, and local programs that build and operate radio systems, and the RapidCom program focuses on the immediate development of incident-response interoperable emergency communications in high-threat urban areas. RapidCom will ensure that high-threat urban areas have incident-level, interoperable emergency communications equipment by September 30, 2004. The program will establish communications interoperability in these urban areas for an incident area approximately the size of the attacks on the World Trade Center towers on September 11th. At the incident area, all emergency personnel from various regional jurisdictions will be able to communicate using existing equipment that is made interoperable by a patch-panel device, interconnecting various models of equipment that would otherwise not be compatible.

Emerging Technologies: Homeland Security's Advanced Research Projects Agency (HSARPA) invests in the private sector, funding revolutionary new technological advances to make America safer. HSARPA has already delivered significant advances in radiological and nuclear detection, biological and chemical countermeasures, and ongoing projects include waterway vessel tracking technology and new cargo security technologies for advanced container security. There have also been great strides made in harnessing scientific advances in biometrics to strengthen travel security and to help detect and counter identity theft. Through the Homeland Security Centers of Excellence program, the Department is creating university-based partnerships to research issues essential to our security, with Centers already established on risk and economic analysis of terrorism, animal disease defense, and food security (and proposals are currently being received for a fourth Center focusing on the sociological and behavioral aspects of terrorism). BioWatch/BioShield: An environmental monitoring system, BioWatch, monitors air samples on a frequent basis in major urban cities nationwide, providing early warning of a potential bio attack which would allow treatment before people get sick. Homeland Security is also deploying and evaluating mobile automatic air testing kits that house biological and chemical sensors for even quicker reporting. This program links the earliest detection possible with efforts to develop medical countermeasures and a program called BioShield that ensures vaccines, drugs and medical supplies are ready for rapid distribution. Integrated Planning: The Department of Homeland Security has led the development of the National Response Plan (NRP), which consolidates and reconciles multiple national-level incident response plans into a single, focused, universally understood strategy. This effort includes the development of a new catastrophic incident response protocol that will greatly accelerate the delivery of critical federal assistance to domestic venues suffering from a mass casualty/mass evacuation incident.

More Money: The 2005 budget request of $40.2 billion for homeland security is $9 billion (29%) over the 2003 level and $20.4 billion over the 2001 level—an increase of 103% over the 2001 level of homeland security funding. Furthermore, from FY 2002-FY 2004 $13.1 billion has been earmarked for first responder and public health terrorism preparedness—an increase of 900% over the $1.2 billion spent in the previous three years. More Training: For FY 2004, Homeland Security has trained 205,480 first responders (451,634 since FY 2002.) Also, DHS initiated the National Incident Management System (NIMS) and established the NIMS Integration Center, which ensures that Federal, state, and local governments and private-sector organizations are all using the same criteria to prepare for, prevent, respond to, and recover from a terrorist attack or other major disaster.

September 7, 2004

Chapter 10   Appendix

2005 BASIC CATALOG OF FEDERAL DOMESTIC ASSISTANCE

Department of Homeland Security

CROSSWALK FOR THE 20
BASIC CATALOG OF FEDERAL DOMESTIC ASSISTANCE (CFDA)

DEPARTMENT OF HOMELAND SECURITY (DHS)
JUNE 2005

Fiscal Year 2004
Fiscal Year 2005

Updated Catalog Number	Basic Catalog Number and/or Title	Remarks
FISCAL YEAR 2003		
97.001	No number	Special Projects	FY 2003 Special programs or projects from various DHS Directorates that are funded by name or earmark under appropriation authorizations and will not be on-going or continued after the fiscal year.
97.002	No number	Research Projects	FY 2003 Research Projects from various DHS Directorates that are funded by name or earmark under appropriation authorizations and will not be on-going or continued after the fiscal year.
97.003	(10.025) CFDA number retained by USDA.	Plant and Animal Disease, Pest Control, and Animal Care	FY 2003, 5 Project transferred from the USDA/APHIS to the DHS/Border and Transportation Directorate. **A new description is necessary because the existing CFDA number will be retained by USDA/APHIS.** **FY 2005 Program deleted from CFDA – no assistance provided to non-Federal entities.**
97.004	16.007	State Domestic Preparedness Equipment Support Program	FY 2003 Program transferred from Dept of Justice/Office of Domestic Preparedness to DHS/Border and Transportation Security Directorate/Office of Domestic Preparedness. FY 2004 program included in the OOSLGCP/ODP 97.004 (State Homeland Security Grant Program) FY 2005 program included in OSLGCP/ODP 97.067 Homeland Security Grant Program. NOTE: The CFDA number is used to award grants independent of the State grants.

97.005	16.008	State and Local Domestic Preparedness Training Program	FY 2003 Program transferred from Dept of Justice/Office of Domestic Preparedness to DHS/Border and Transportation Security Directorate/Office of Domestic Preparedness. FY 2004 program included in the OSLGCP/ODP 97.004 (State Homeland Security Grant Program) FY 2005 program included in OSLGCP/ODP 97.067 Homeland Security Grant Program. NOTE: The CFDA number is used to award grants independent of the State grants.
97.006	16.009	State and Local Domestic Preparedness Exercise Support	FY 2003 Program transferred from Dept of Justice/Office of Domestic Preparedness to DHS/Border and Transportation Security Directorate/Office of Domestic Preparedness. FY 2004 program included in the OSLGCP/ODP 97.004 (State Homeland Security Grant Program) FY 2005 program included in OSLGCP/ODP 97.067 Homeland Security Grant Program. NOTE: The CFDA number is used to award grants independent of the State grants.
97.007	16.010	State and Local Domestic Preparedness Technical Assistance	FY 2003 Program transferred from Dept of Justice/Office of Domestic Preparedness to DHS/Border and Transportation Security Directorate/Office of Domestic Preparedness. NOTE: The CFDA number is used to award grants independent of the State grants.
97.008	16.011	Urban Areas Security Initiative	FY 2003 Program transferred from Dept of Justice/Office of Domestic Preparedness to DHS/Border and Transportation Security Directorate/Office of Domestic Preparedness. FY 2005 program included in 97.067 Homeland Security Grant Program.
97.009	16.201	Cuban and Haitian Entrant Resettlement Program	FY 2003 Adult Program transferred from Department of Justice/Office of International Affairs/Immigration and Naturalization Service to DHS/Border and Transportation Security Directorate.
97.010	16.400	Citizenship Education and Training	FY 2003 Program transferred to the DHS/Bureau of Immigration and Customs Enforcement/Office of Citizenship from the Department of Justice.

97.011	20.001	Boating Safety	FY 2003 Program transferred to the DHS/Commandant of Coast Guard through the transfer of the U.S. Coast Guard from the Department of Transportation.
97.012	20.005	Boating Safety Financial Assistance	FY 2003 Program transferred to the DHS/Commandant of Coast Guard through the transfer of the U.S. Coast Guard from the Department of Transportation.
97.013	20.006	State Access to the Oil Spill Liability Trust Fund	FY 2003 Program transferred to the DHS/Commandant of Coast Guard through the transfer of the U.S. Coast Guard from the Department of Transportation.
97.014	20.007	Bridge Alteration	FY 2003 Program transferred to the DHS/Commandant of Coast Guard through the transfer of the U.S. Coast Guard from the Department of Transportation.
97.015	21.100	Secret Service Training Activities	FY 2003 Program transferred to the DHS/Secret Service on with the incorporation of the Federal Emergency Management Agency into the DHS.
97.016	83.007	Reimbursement for Firefighting on Federal Property	FY 2003 Program transferred to the DHS/Emergency Preparedness and Response Directorate/U.S. Fire Administration with the incorporation of the Federal Emergency Management Agency into the DHS.
97.017	New Program	Pre-Disaster Competitive Program	FY 2003 New Program
97.018	83.009	National Fire Academy Training Assistance	FY 2003 Program transferred to the DHS/Emergency Preparedness and Response Directorate/U.S. Fire Administration with the incorporation of the Federal Emergency Management Agency into the DHS.
97.019	83.010	National Fire Academy Educational Program	FY 2003 Program transferred to the DHS/Emergency Preparedness and Response Directorate/U.S. Fire Administration with the incorporation of the Federal Emergency Management Agency into the DHS.
97.020	83.011	Hazardous Materials Training Program	FY 2003 Program transferred to the DHS/Emergency Preparedness and Response Directorate/Office of National Preparedness with the incorporation of the Federal Emergency Management Agency into the DHS.
97.021	83.012	Hazardous Materials Assistance	FY 2003 Program transferred to the DHS/Emergency Preparedness and Response Directorate/Office of National Preparedness with the incorporation of the Federal Emergency Management Agency into the DHS.
97.022	83.100	Flood Insurance	FY 2003 Program transferred to the DHS/Emergency Preparedness and Response Directorate/Federal Insurance and Mitigation Administration with the incorporation of the Federal Emergency Management Agency into the DHS.

97.023	83.105	Community Assistance Program State Support Services Element (CAP-SSSE)	FY 2003 Program transferred to the DHS/Emergency Preparedness and Response Directorate/Federal Insurance and Mitigation Administration with the incorporation of the Federal Emergency Management Agency into the DHS.
97.024	83.523	Emergency Food and Shelter National Board Program	FY 2003 Program transferred to the DHS/Emergency Preparedness and Response Directorate/ Response and Recovery Directorate with the incorporation of the Federal Emergency Management Agency into the DHS.
97.025	83.526	National Urban Search and Rescue (US&R) Response System	FY 2003 Program transferred to the DHS/Emergency Preparedness and Response Directorate/ Response and Recovery Directorate with the incorporation of the Federal Emergency Management Agency into the DHS.
97.026	83.527	Emergency Management Institute-Training Assistance (Student Stipend Reimbursement Program (SEP)	FY 2003 Program transferred to the DHS/Emergency Preparedness and Response Directorate/ Response and Recovery Directorate with the incorporation of the Federal Emergency Management Agency into the DHS.
97.027	83.529	Emergency Management Institute (EMI)_Independent Study Program	FY 2003 Program transferred to the DHS/Emergency Preparedness and Response Directorate/ Response and Recovery Directorate with the incorporation of the Federal Emergency Management Agency into the DHS.
97.028	83.530	Emergency Management Institute (EMI)_Resident Educational Program	FY 2003 Program transferred to the DHS/Emergency Preparedness and Response Directorate/ Response and Recovery Directorate with the incorporation of the Federal Emergency Management Agency into the DHS.
97.029	83.536	Flood Mitigation Assistance	FY 2003 Program transferred to the DHS/Emergency Preparedness and Response Directorate/Federal Insurance and Mitigation Administration with the incorporation of the Federal Emergency Management Agency into the DHS.
97.030	83.537	Community Disaster Loans	FY 2003 Program transferred to the DHS/Emergency Preparedness and Response Directorate/ Response and Recovery Directorate with the incorporation of the Federal Emergency Management Agency into the DHS.
97.031	83.538	Cora Brown Fund	FY 2003 Program transferred to the DHS/Emergency Preparedness and Response Directorate/ Response and Recovery Directorate with the incorporation of the Federal Emergency Management Agency into the DHS.
97.032	83.539	Crisis Counseling	FY 2003 Program transferred to the DHS/Emergency Preparedness and Response Directorate/ Response and Recovery Directorate with the incorporation of the Federal Emergency Management Agency into the DHS.

97.033	83.540	Disaster Legal Services	FY 2003 Program transferred to the DHS/Emergency Preparedness and Response Directorate/ Response and Recovery Directorate with the incorporation of the Federal Emergency Management Agency into the DHS.
97.034	83.541	Disaster Unemployment Assistance	FY 2003 Program transferred to the DHS/Emergency Preparedness and Response Directorate/ Response and Recovery Directorate with the incorporation of the Federal Emergency Management Agency into the DHS.
97.035	83.543	Individual and Family Grants	FY 2000 Program rescinded under Disaster Mitigation Act of 2000, but CFDA retained for audit purposes.
			FY 2003 Program transferred to the DHS/Emergency Preparedness and Response Directorate/ Response and Recovery Directorate with the incorporation of the Federal Emergency Management Agency into the DHS.
97.036	83.544	Public Assistance Grants	FY 2003 Program transferred to the DHS/Emergency Preparedness and Response Directorate/ Response and Recovery Directorate with the incorporation of the Federal Emergency Management Agency into the DHS.
97.037	83.545	Disaster Housing Program	FY 2000 Program rescinded under Disaster Mitigation Act of 2000, but CFDA retained for audit purposes.
			FY 2003 Program transferred to the DHS/Emergency Preparedness and Response Directorate/ Response and Recovery Directorate with the incorporation of the Federal Emergency Management Agency into the DHS.
97.038	83.547	First Responder Counter-Terrorism Training Assistance (Counter-Terrorism Training (ATT))	FY 2003 Program transferred to the DHS/Emergency Preparedness and Response Directorate/U.S. Fire Administration with the incorporation of the Federal Emergency Management Agency into the DHS.
97.039	83.548	Hazard Mitigation Grant	FY 2003 Program transferred to the DHS/Emergency Preparedness and Response Directorate/Federal Insurance and Mitigation Administration with the incorporation of the Federal Emergency Management Agency into the DHS.
97.040	83.549	Chemical Stockpile Emergency Preparedness Program	FY 2003 Program transferred to the DHS/Emergency Preparedness and Response Directorate/Office of National Preparedness with the incorporation of the Federal Emergency Management Agency into the DHS.

97.041	83.550	National Dam Safety Program (Dam Safety State Assistance Program)	FY 2003 Program transferred to the DHS/Emergency Preparedness and Response Directorate/Federal Insurance and Mitigation Administration with the incorporation of the Federal Emergency Management Agency into the DHS.
97.042	83.552	Emergency Management Performance Grants	FY 2003 Program transferred to the DHS/Emergency Preparedness and Response Directorate/Office of National Preparedness with the incorporation of the Federal Emergency Management Agency into the DHS. FY 2004 Transferred to OSLGCP. FY 2005 Incorporated into consolidated grant application and award under 97.067 Homeland Security Grant Program.
97.043	83.553	State Fire Training Systems Grants (National Fire Academy Training Grants)	FY 2003 Program transferred to the DHS/Emergency Preparedness and Response Directorate/U.S. Fire Administration with the incorporation of the Federal Emergency Management Agency into the DHS.
97.044	83.554	Assistance to Firefighters Grant (Fire Grants)	FY 2003 Program transferred to the DHS/Emergency Preparedness and Response Directorate/U.S. Fire Administration with the incorporation of the Federal Emergency Management Agency into the DHS. FY 2004 program transferred to OSLGCP from EP&R.
97.045	83.555	Cooperating Technical Partners	FY 2003 Program transferred to the DHS/Emergency Preparedness and Response Directorate/Federal Insurance and Mitigation Administration with the incorporation of the Federal Emergency Management Agency into the DHS.
97.046	83.556	Fire Management Assistance Grant	FY 2003 Program transferred to the DHS/Emergency Preparedness and Response Directorate/ Response and Recovery Directorate with the incorporation of the Federal Emergency Management Agency into the DHS.
97.047	83.557	Pre-Disaster Mitigation	FY 2003 Program transferred to the DHS/Emergency Preparedness and Response Directorate/Federal Insurance and Mitigation Administration with the incorporation of the Federal Emergency Management Agency into the DHS.

97.048	83.558	Federal Assistance to Individuals and Households Housing	FY 2003 Program transferred to the DHS/Emergency Preparedness and Response Directorate/ Response and Recovery Directorate with the incorporation of the Federal Emergency Management Agency into the DHS.
97.049	83.559	Federal Assistance to Individuals and Household Disaster Housing Operations	FY 2003 Program transferred to the DHS/Emergency Preparedness and Response Directorate/ Response and Recovery Directorate with the incorporation of the Federal Emergency Management Agency into the DHS.
97.050	83.560	Federal Assistance to Individuals and Household – Other Needs (Individual and Household Other Needs)	FY 2003 Program transferred to the DHS/Emergency Preparedness and Response Directorate/ Response and Recovery Directorate with the incorporation of the Federal Emergency Management Agency into the DHS.
97.051	83.562	State and Local All Hazards Emergency Operations Planning (S/L Emergency OPS Planning)	FY 2003 Program transferred to the DHS/Emergency Preparedness and Response Directorate/Office of National Preparedness with the incorporation of the Federal Emergency Management Agency into the DHS. One year funding/program discontinued.
97.052	83.563	Emergency Operations Centers	FY 2003 Program transferred to the DHS/Emergency Preparedness and Response Directorate/Office of National Preparedness with the incorporation of the Federal Emergency Management Agency into the DHS. Once year funding/program discontinued.
97.053	83.564	Citizen Corps	FY 2003 Program transferred to the DHS/Emergency Preparedness and Response Directorate/Office of National Preparedness with the incorporation of the Federal Emergency Management Agency into the DHS. FY 2004 program transferred to OSLGCP from EP&R and included in 97.004 (State Homeland Security Grant Program) FY 2005 program included in 97.067 Homeland Security Grant Program.
97.054	83.565	Community Emergency Response Teams (CERT)	FY 2003 Program transferred to the DHS/Emergency Preparedness and Response Directorate/Office of National Preparedness with the incorporation of the Federal Emergency Management Agency into the DHS. FY 2004 program transferred to OSLGCP from EP&R

Number	Program	Notes
97.055	Interoperable Communications Equipment	FY 2003 Program transferred to the DHS/Emergency Preparedness and Response Directorate/Office of National Preparedness with the incorporation of the Federal Emergency Management Agency into the DHS.
83.566		FY 2004 program transferred to OSLGCP from EP&R.
97.056	Port Security Grant Program	FY 2003 Program transferred from the Department of Transportation to the DHS/Border and Transportation Security/Transportation Security Administration.
New Program		FY 2004 Program transferred to OSLGCP from EP&R.
97.057	Intercity Bus Security Grants	FY 2003 Program transferred from the Department of Transportation to the DHS/Border and Transportation Security/Transportation Security Administration.
New Program		FY 2004 Program transferred to OSLGCP from EP&R.
97.058	Operation Safe Commerce	FY 2003 Program transferred from the Department of Transportation to the DHS/Border and Transportation Security/Transportation Security Administration.
New Program		FY 2004 Program transferred to OSLGCP from EP&R.
97.059	Truck Security	FY 2004 New program from Border and Transportation Security/Transportation Security Administration.
New Program		FY 2004 Program transferred to OSLGCP from EP&R.
97.060	Port Security Research and Development Grant	FY 2003 program transferred from DOT/Transportation Security Administered; administered by DOT/FAA.
New Program		
97.061	Centers for Homeland Security	FY 2003 New program for Science & Technology Directorate.
New Program		
97.062	Scholars and Fellows	FY 2003 New program for Science & Technology Directorate.
New Program		
FISCAL YEAR 2004		
97.063	Pre-Disaster Mitigation Disaster Resistance University	FY 2004 New program for EP&R.
New Program		FY 2005 Program incorporated in to 97.017.
97.064	Debris Removal Insurance	FY 2004 New program for EP&R
New Program		
97.065	Homeland Security Advance Research Projects Agency	FY 2003 New program for Science & Technology Directorate
New Program		
97.066	Homeland Security Information Technology and Evaluation	FY 2004 New program for CIO, implemented by OSLGCP, awarded under DOJ CFDA 16.000.
16.000		

Number	Status	Program	Description
97.067	New Program	Homeland Security Grant Program	FY 2004 New program for OSLGCP to incorporate several programs into a single application and grant award.
		97.004 State Domestic Preparedness Equipment Support Program	FY 2003 Incorporated into 97.004 (State Homeland Security Program); FY 2004 program included in 97.004 (State Homeland Security Grant Program);FY 2005 program included in 97.067 Homeland Security Grant Program.
		97.005 State & Local Domestic Preparedness Training Program	FY 2003 Incorporated into 97.004 (State Homeland Security Program); FY 2004 program included in 97.004 (State Homeland Security Grant Program); FY 2005 program included in 97.067 Homeland Security Grant Program.
		97.006 State & Local Domestic Preparedness Exercise Support	FY 2003 Incorporated into 97.004 (State Homeland Security Program); FY 2004 program included in 97.004 (State Homeland Security Grant Program); FY 2005 program included in 97.067 Homeland Security Grant Program.
		97.053 Citizen Corps	FY 2004 Incorporated into 97.067 Homeland Security Grant Program; FY 2005 program included in 97.067 Homeland Security Grant Program.
		Law Enforcement Terrorism Prevention	FY 2004 Incorporated into 97.067 Homeland Security Grant Program; FY 2005 program included in 97.067 Homeland Security Grant Program.
		97.008 Urban Areas Security Initiative	FY 2005 Incorporated into 97.067 Homeland Security Grant Program.
		97.042 Emergency Management Performance Grants	FY 2005 Incorporated into 97.067 Homeland Security Grant Program.
		97.071 Metropolitan Medical Response System	FY 2005 Incorporated into 97.067 Homeland Security Grant Program.
97.068	New Program	Competitive Training Grants	FY 2004 New program for OSLGCP.
97.069	New CFDA	Aviation Research Grants	FY 2004 Existing Program no CFDA number
97.070	New Program	Map Modernization Management Support	FY 2004 New program for EP&R

CFDA	Status	Program	Notes
97.071	New CFDA	Metropolitan Medical Response System	FY 2003 Program transferred from HHS to EP&R with no CFDA number. FY 2004 Transferred to OSLGCP from EP&R. FY 2005 Incorporated into 97.067 Homeland Security Grant Program.
97.072	New CFDA	National Explosive Detection Canine Team Program	FY 2004 Existing TSA program assigned CFDA number.
97.073	New CFDA	State Homeland Security Program (SHSP)	FY 2004 Program cross-walk from 97.004
97.074	New CFDA	Law Enforcement Terrorism Prevention Program (LETPP)	FY 2004 New OSLGCP Program.
97.075	New Program	Rail and Transit Security Grant Program	FY 2004 New OSLGCP Program.
FISCAL YEAR 2005			
97.076	New CFDA	National Center for Missing and Exploited Children	FY 2005 Existing USSS program assigned CFDA number
97.077	New Program	Homeland Testing, Evaluation and Demonstration of Technologies	FY 2005 New program for S & T
97.078	New Program	Buffer Zone Protection (BZP)	FY 2005 New program for IAIP; implemented by OSLGCP.
97.079	New Program	Public Radios for Schools	FY 2005 New program for IAIP, implemented by NOAA
97.080	New Program	IAIP Pilot Projects	FY 2005 New program for IAIP
97.081	New CFDA	Law Enforcement Training and Technical Assistance	FY 2005 Existing program for FLETC assigned CFDA number.
97.082	New CFDA	Earthquake Consortium	FY 2005 Existing program for EP&T/FEMA assigned CFDA.
97.083	New Program	Staffing for Adequate Fire Emergency Response (SAFER)	FY 2005 New program for OSLGCP.

INDEX

Page numbers in italics refer to figures